Laboratory Exercises in
OCEANOGRAPHY

Third Edition

Bernard W. Pipkin
University of Southern California

Dean A. Dunn
University of Southern Mississippi

Donn S. Gorsline
University of Southern California

Stephen A. Schellenberg
San Diego State University

Richard E. Casey
University of San Diego

D0218451

 W. H. Freeman and Company / New York

About the cover:
Data for this oblique false-color image of seafloor features off of California was gathered by extensive high-resolution, multi-beam bathymetric studies in the mid-1980s to early 1990s by the National Oceanographic and Atmospheric Administration. The studies were commissioned by the National Science Foundation after the Reagan Administration declared a 323-kilometer (200-mile) band of U.S. coastal waters off limits to foreign fishing and seafloor mining in 1983. Dr. Lincoln F Pratson (University of Colorado) and Dr. William F. Haxby (Columbia University) analyzed the raw data to produce the oblique, false-color image that provides an insightful perspective on the California seafloor.

The continental shelf along coastal California, shown in shades of grey (less than ~150) meters), is especially wide near San Francisco. The continental slope, shown in shades of pink to yellow (~150–2,000 meters), slopes seaward at various degrees, and grades into the continental rise, shown in shades of green to light blue (~2,000–3,000 meters). The continental rise merges with the abyssal plain, shown in dark blue (~3,000+ meters). Numerous submarine canyons are incised through the continental shelf and slope, with some extending as far and deep as the abyssal plain. The rumbled textures on the continental rise and abyssal plains are extensive undersea fans of sediment that have been shed off the continental shelf and slope. The abyssal plain is occasionally interrupted by seamounts, such as the Davidson Seamount in the lower-left of the image. All of these geographic features lie on the Pacific Plate, with the San Andreas and associated transform faults in the east forming the boundary between the Pacific and North American Plates. This complex shearing along the plate boundary appears to be causing the lower half of the continental slope to bulge outward from the continent, forming an extensive mid-depth plateau in the foreground. High-resolution bathymetric data were not collected for the region west of the black line along the left margin.

Similar maps of the other U. S. Continental margin areas are available at <www.ldeo.columbia.edu/datarep/index.html> and in Pratson and Haxbys research papers in the journals *Scientific American* (June 1997) and *Geology* (January 1996).

Cover image: Lincoln P Pratson, University of Colorado, and William F. Haxby, Columbia University, "Panoramas of the Seafloor," *Scientific American,* June 1997, pp. 82-83.

ISBN-13: 978-0-7167-3742-1
ISBN-10: 0-7167-3742-6

Printed in the United States of America

Contents

Exercises

Preface

In the prefaces to the first and second editions of this manual we reviewed the importance of the oceans to human culture and economic activities. This importance has increased with the growing concern about the global changes that can be effected by human activities, such as our contributions of "greenhouse gases" to the atmosphere. Carbon dioxide is produced by the combustion of fossil fuels. Methane is generated by cattle and from degradation of organic matter under bacterial attack in oxygen-deficient environments. These additions to natural sources plus the universal presence of water vapor can create significant changes in the oceans, since gases diffuse into and out of the oceans in response to changes in the overlying atmosphere. The role of the oceans in moderating global climate changes is a major area for research, and many of the mechanisms are still not well known.

In the years since we first initiated our elementary course in oceanography at the University of Southern California, a number of significant problems have become more widely appreciated. Not the least of these is the increasing pressure on many ocean fisheries. Many of the traditional major fishery areas are now in critical condition. As human populations have grown, increased pressure on the coastal environment has resulted from development of new residential and recreational areas and increasing activity in all major ports. These pressures have economic impacts that are felt worldwide. Much of the exploration and exploitation of energy resources is now concentrated in offshore areas, and the transfer of oil and gas to shore and ports is a major economic activity. Environmental concerns also arise from this transfer. Thus it is important that the citizens of our country and all coastal nations have a general understanding of the characteristics, dimensions, and processes active in the oceans.

Our knowledge of the oceans has continued to increase as vast amounts of data pour in from deep ocean drilling, global oceanographic circulation studies, and studies of air-sea exchanges on a global scale. We are discovering that the dynamics of ocean and atmosphere are the product of a spectrum of cyclic phenomena that range in period from seasonal cycles to long-term cycles hundreds of thousands of years. Some changes can be surprisingly rapid. As the warming following the last glacial maximum developed beginning about 17,000 years ago, several collateral effects, including the release of large amounts of meltwater and changes in the albedo of large areas, created a reversal to colder conditions about 11,000 years ago. This shift may have occurred in as little as 30 years or less, well within a single human lifetime. As we look at the characteristics of deep-ocean sediments spanning the last few hundreds of thousands of years and compare those data with information from ice cores in the Arctic and Antarctic, and lake records from many areas, we see numerous cyclic changes. Archaeologists and anthropologists are beginning to relate some of these to significant episodes in human cultural development. The rise of agriculture may have been associated with some of the changes noted above, and shifts in population, the filling of ancient harbors, and changes in sea level both positive and negative in given areas all had influences on early human societies.

This third edition has been completely reviewed, errors corrected, and many exercises revised. Some exercises have been dropped on the basis of our own and other instructors' experience, some combined, and some newly prepared in response to suggestions by many teachers, whose combined experience has been of immeasurable value. Thousands of students have passed through our course, and they have contributed to the changes in this edition, either by reactions to some laboratory exercises in class, or by explicit comments by some highly motivated students. All of their assistance and advice is highly appreciated.

We know that many elementary oceanography classes do not have a formal laboratory, but we suggest that the manual provides much useful material that can be used for homework assignments, or as supplements to lecture material. We have provided more exercises than can be accomplished in a single semester or quarter in order to provide flexibility for the instructor in selecting a set of exercises that highlight the major topics of a given course. The exercises can be done with minimal material and equipment and can be supplemented by equipment, videos, transparencies, and demonstrations. Students can also augment the laboratory work with material collected on local field excursions.

We would like to gratefully acknowledge the insightful reviews that added substantially to the form and substance of this edition. Particular thanks go to Beth A. Christensen of Furman University; Steve P. Lund, University of Southern California; Steve Macko, University of Virginia; Carol J. Pride, Marine Science Institute, University of California at Santa Barbara; and Russell Scott Shapiro of the College of the Redwoods in Eureka, California.

Our editors at W. H. Freeman and Company provided encouragement and helpful criticism throughout the preparation and publication of the third edition, for which we are most grateful. We are indebted to Diane Davis, our Project Editor, who performed a miracle in transforming a five-author cut-and-paste rough manuscript into publishable form; to Jennifer MacMillan for her photo research skills; and to Brian Donnellan for sorting out and keeping track of the figures. We appreciate the encouragement and direction given us by Melissa Wallerstein, who was our Science Editor through most of the project.

Bernard W. Pipkin
Donn S. Gorsline
Richard E. Casey
Dean A. Dunn
Stephen A. Schellenberg

Introduction

The interpretation and application of new observations and measurements of ocean characteristics and processes requires laboratory analysis and data classification. These methods are drawn from the spectrum of basic sciences, including physics, chemistry, biology, and geology. In the majority of the exercises in this manual the basic information is real data collected from the world oceans by workers on many ships and other platforms using a variety of sampling methods. Satellite data are increasingly important in ocean study, and the results of some of these observations are also a part of the background information.

The primary purpose of these exercises is to give you practice and experience in manipulation, evaluation, and interpretation of ocean data. Each exercise is designed to extend and complement some topic in the lecture materials. A more general purpose is to develop an appreciation of the methods and approaches used in all scientific studies. The exercises will also give you an impression of the scale of features, events, and processes that operate in the world ocean, and of the geologic processes and characteristics that affect the size, shape, and evolution of the ocean basins. Input from the land is evaluated in the study of ocean sedimentary materials and in measurements of rates of accumulation of these sediments in different ocean environments. Every major aspect of the oceans is touched upon, including studies of the biological communities, processes, and interactions that are part of the ocean's productivity. We also deal with chemical and physical aspects of the oceans. No one course will use all of the exercises, but your instructor will select those that highlight the points of emphasis in the framework of your course. Each exercise includes discussions of the background information on the topic, introduces new terms and their definitions, and provides forms for data tabulation and calculation. Those who may be interested in delving more deeply into certain topics will find additional readings listed in the Bibliography.

All of the exercises will require the manipulation of data and plotting of curves and maps. These experiences will introduce you to the mathematical manipulation of data and the best ways of presenting data to get at answers to the major questions. You will also become familiar with rates and scales, dimensions and conversions, and a number of other techniques for analyzing and displaying data to reach conclusions. Classification is also a basic method for sorting and visualizing data. None of the operations in these exercises are very demanding, and can easily be accomplished by all students with a high school background in science and mathematics. However, we hope that the completion of these exercises in association with the lecture material will give you a sound understanding of how science works and how problems can be approached and solved quantitatively. We also hope that you will leave the course with a much more comprehensive understanding of the world ocean.

Bathymetry—the Shape of the Sea Floor

OBJECTIVES:

■ To visualize a three-dimensional sea-floor surface from a two-dimensional construct (marine chart or map).

■ To become familiar with the mechanics and rules of contouring charts from an array of individual sea-floor depth measurements (soundings).

■ To reveal the shape of the sea floor or specific sea-floor features by constructing topographic profiles.

We are all aware of the variety of landforms to be seen on earth, from colossal mountain ranges to deep canyons. However, many more forms, indeed the majority of geographic features on the entire earth, are covered by water. Although direct observation and measurement of these forms are impossible, what information we do have reveals that the most dramatic marvels on land are exceeded by others on the sea floor: there are, for instance, submarine mountain ranges longer and higher than the Rockies, and magnificent canyons into which the Grand Canyon would fit with room to spare.

Bathymetry

The term *bathymetry*, from the Greek roots *bathy-* meaning depth, and *-metry* meaning the process of measurement, is defined as the measuring and charting of the topography of the sea floor. At one time **soundings** (depth measurements) were accomplished by dropping a weighted line from a vessel, a task that often took several hours in deep

water. Today a more practical electronic device, developed by the U.S. Navy in 1922, enables us to collect a sufficient number of accurately located depths, which in turn disclose details of sea-floor topography. The device, known as an *echo sounder,* or *fathometer* (1 fathom = 6 feet), records the time required for a sound pulse to travel from the ship to the sea floor and back again. Knowing the velocity of sound in seawater (about 4800 feet, or about 1450 meters, per second), we can easily calculate depth if the travel time is known (Figure 1-1). In fact, the fathometer is simply a very accurate electronic timer that records the travel times of reflected sound pulses and marks these "echoes" on a graph, or *strip-chart recorder,* at the appropriate depth in feet, meters, or fathoms. The resulting display on the strip chart is known as an *echogram* (Figure 1-2).

In practice, soundings are recorded continuously by the instrument. Periodic notations of time of day, ship speed, and direction of travel are written by the investigator directly on the recording paper, or strip chart. Later the soundings are marked on a chart or base map, the position of each sound being determined by the navigator's chart and the notations on the strip chart. By this procedure a great number of reliable soundings from many different ships have been compiled to construct accurate charts for navigational purposes and scientific analysis. **Contour** lines, called **isobaths** in marine work, may be constructed after soundings have been placed on the chart. The shape and spacing of the isobaths will reveal the features of that area of the ocean floor.

A chart or map must also have a **scale.** This can be a calibrated bar that shows the relationship between the units of length used on the particular

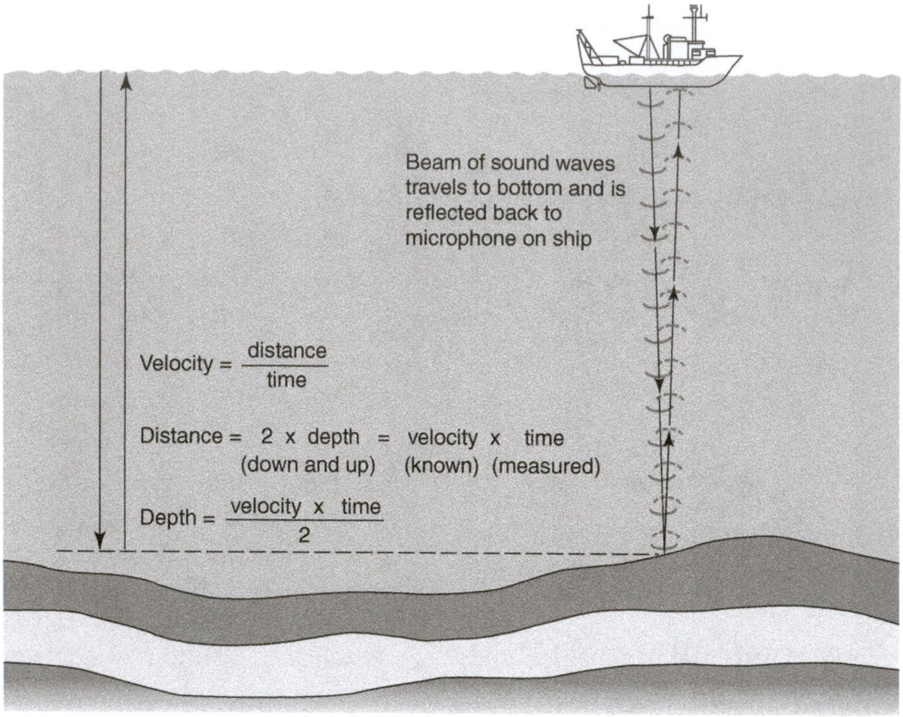

Beam of sound waves
travels to bottom and is
reflected back to
microphone on ship

$$Velocity = \frac{distance}{time}$$

$$Distance = 2 \times depth = velocity \times time$$
$$\qquad\quad (down\ and\ up) \quad (known) \quad (measured)$$

$$Depth = \frac{velocity \times time}{2}$$

Figure 1-1 Echo sounders sense underwater topography by beaming sound waves to the bottom and measuring the time required for the beam to be reflected back to the ship. [From F. Press and R. Siever, *Earth,* 4th ed. W. H. Freeman and Company. Copyright © 1986.]

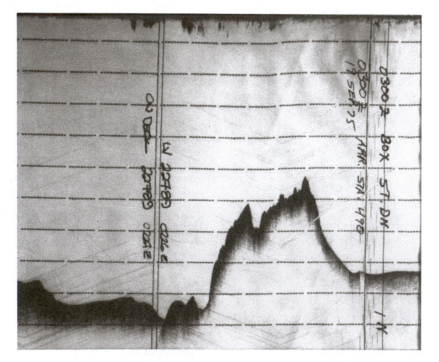

Figure 1-2 Echogram of Lasuen Knoll due west of Dana Point, California. Horizontal lines are depth intervals of 20 fathoms; vertical lines with written notations are locations at which bottom sampling was performed. The depth to the top of the seaknoll in this crossing is about 105 fathoms (191 meters), and the bottom materials at Station 490 were fine sand and rock. The top line is the sea surface.

map and the actual units they represent, such as nautical miles or kilometers. The scale may also be a fractional one: for example, a scale of 1:500,000 means that one unit on the map (centimeters, millimeters, or inches) is equal to 500,000 of the same units. Thus 1 centimeter on the map is equal to 500,000 centimeters true distance on the earth, which is equal to 5000 meters or 5 kilometers.

The Construction of a Bathymetric Chart

In constructing a bathymetric chart from plotted soundings, observe the following simple rules:

1. Isobaths never cross one another.

2. When isobaths cross a canyon or deeply incised valley, they have a V-shape with the apex of the V pointing upstream (toward shallower water).

3. An isobath must separate depth zones completely; that is, if a 100-meter isobath is drawn off-shore along a coastline, all depths of 100 meters must be on the line, and all soundings shallower than 100 meters must be between the isobath and the shore. The same principle applies to constructing deeper lines: the area between the 100- and 200-meter lines must not contain soundings less than 100 meters or greater than 200 meters.

4. Closely spaced isobaths represent a steep slope or an abrupt change in depth. Widely spaced isobaths represent a gentle slope or gradual change in depth.

5. An isobath closes upon itself. The smaller features may close on the chart, but larger features may run off the chart as in Figure 1-3b.

Figure 1-3a is a perspective view of two submarine hills separated by a valley. The hill on the right is low and rounded, extending from 900 meters to 250 meters below sea level. The hill on the left side consists of two peaks with a steep east-facing slope and numerous gullies on the lower slopes. Figure 1-3b is a map view bathymetric chart of the same features.

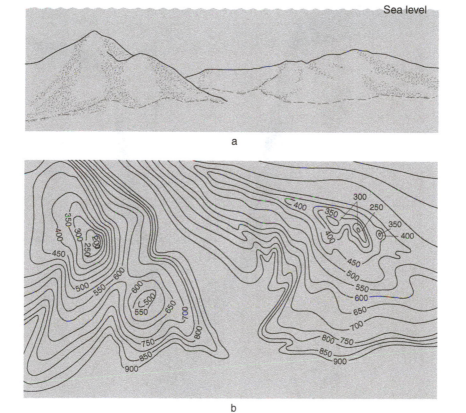

Figure 1-3 Representation of an area of sea floor with two hills separated by a valley. Part (a) is a perspective view; part (b) is a map view in which the features in (a) are represented by isobaths. Depths noted along the isobaths are in meters.

The most accurate device for measuring and charting submarine topography is a precision depth recorder (PDR). It is similar to a fathometer in construction and placement aboard ship except that the timing of sound pulses is precisely controlled by use of a tuning fork within the instrument. Instruments with timing accuracies of better than one part in 3000 have been developed in the United States and Europe. Such precision is not required for routine navigation purposes but is mandatory for accurate scientific work. Figure 1-4 is part of a PDR profile of the Redondo Submarine Canyon. Vertical lines, such as the two shown on the figure, are marked every 15 minutes as well as whenever there is a change in the ship's course. On this recording the navigator on the bridge fixed the ship's position at 1200 hours and 1209 hours so that the scientist could accurately locate the axis of the submarine canyon, noted at 1203 hours, on a marine chart. Also shown are the ship's course (C 175°) and its speed of 7 knots (S7K). The regularly spaced horizontal lines are 10-fathom (18-meter) depth intervals, and they indicate that the north edge of the canyon is at a depth of about 100 meters. This portion of the canyon was recorded on a 0-to-200 fathoms scale; however, as the vessel proceeds farther offshore, where the canyon progressively deepens, the scale must be changed to 200–400 fathoms, 400–600 fathoms, and so on, to accommodate the greater depth.

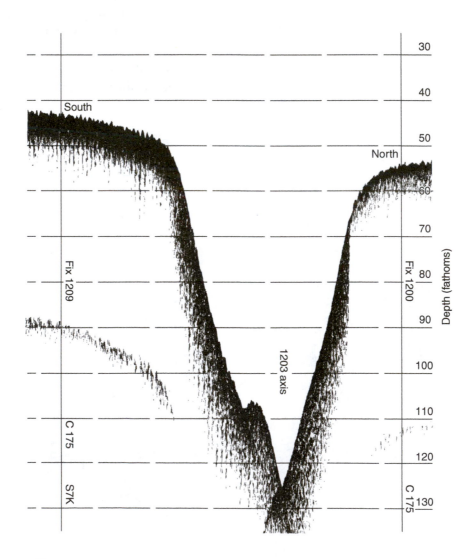

Figure 1-4 Echogram (PDR) of Redondo Submarine Canyon off Redondo Beach, California. Horizontal lines at 10-fathom (18-meter) intervals.

Topographic Cross Sections

A bathymetric chart shows a visualization of the topography of the sea floor as viewed from above. We can derive topography from a contour map or nautical chart by constructing a topographic profile along a preselected line or course. The profile is like a graph with sea-floor depths plotted vertically and distance plotted horizontally. The echograms of Figures 1-2 and 1-4 are topographic profiles on which the data have been plotted electronically.

Commonly, in order to show surface topography in its true relationship to horizontal distance, a profile has the same vertical and horizontal scale. At sea, however, distances are so great relative to **relief** that we must exaggerate the vertical scale. That is, 1 kilometer on the vertical scale of the graph may be five times the length of 1 kilometer on the horizontal scale. In this case the vertical exaggeration (V. E.) is said to be five times (5×). To determine the vertical exaggeration, divide the distance represented by one unit on the horizontal scale by the distance represented by one unit on the vertical grid. For instance, if the horizontal scale is 1:250,000 and the vertical scale is 1:10,000, the vertical exaggeration is

$$\text{V. E.} = \frac{250,000 \text{ (distance of one horizontal unit)}}{10,000 \text{ (distance of one vertical unit)}}$$

$$= 25\times$$

To construct a cross section, choose a profile line and lay a piece of paper along that line. Mark the ends of the profile line on the paper, and wherever an isobath intersects the paper, make a pencil tick and write in the depth of that contour. When this step is completed, lay the paper along the base of a graph or grid that is scaled off vertically for depth and horizontally for distance, and make a dot at the depth on the vertical scale that corresponds to each horizontal distance. Connect the points with a smooth curve to complete the profile. If you are exaggerating the vertical scale, indicate this fact. Although in this exercise the lines of section and profile graph are provided, you are not restricted by them. Should you prefer, you may make a profile of any features shown on either chart with a pencil and graph paper as described.

DEFINITIONS

Contour. A line on a chart, map, or section that connects equal values of a given dimension. A *contour interval* is the vertical distance between adjacent contour lines.

Isobath. A line of equal depth on a chart, equivalent to a contour line on land maps.

Relief. Vertical distance between the highest and lowest points in an area. High relief describes generally rugged topography; low relief describes flat or monotonous terrain.

Scale. A calibrated line or bar that shows the relationship between the units of length used on a particular chart or map and the actual units they represent. The scale may also be a fractional one: for example, a scale of 1:500,000 means that one unit on the map (centimeters, millimeters, or inches) is equal to 500,000 of the same units on the earth.

Sounding. A measurement of water depth.

Exercise 1

Bathymetry — the Shape of the Sea Floor

NAME _____

DATE _____

INSTRUCTOR _____

Your instructor will indicate which chart or charts you should work with.

1. Figure 1-5 on the following page is part of NOAA Chart 411, Gulf of Mexico. It has been reduced in size, but the distance between latitudes 26°–28°N is about 190 kilometers (92 nautical miles) and the distance between longitudes 84°–90°W is 450 kilometers (250 nautical miles). Note the delta of the Mississippi River and the city of New Orleans, Louisiana, in the northwestern quadrant of the map. The continental shelf is the gently sloping (1°–2°) area that extends from the shoreline to the edge of the continental slope (4°–10°) at 50 fathoms. The soundings on the map are in fathoms (1 fathom = 6 feet). Contour the soundings using a 20-fathom contour interval to a depth of 40 fathoms, then contour soundings at depths of 50, 100, 200, 500, and 1000 fathoms.

(a) Does the 50-fathom isobath allow you to discriminate between the Gulf of Mexico Shelf off Louisiana, Mississippi, and Alabama, compared to the width of the West Florida Shelf? _____
Which shelf is wider? _____

(b) The northern Gulf of Mexico is a great pile of Mississippi River sediment, whereas the West Florida Shelf has no major river or sediment contributor. In the absence of sand and silt from rivers, what do you think might have built this wide shelf? The water is warm and provides a favorable environment for marine animals. _____
_____
_____
_____
_____

(c) Label and mark with a dashed line the axis of the Desoto Submarine Canyon, the largest canyon seen on the chart and located off Pensacola Bay. Which conveys a better picture of sea-floor topography, individual soundings or contoured isobaths? Why? _____
_____
_____
_____
_____

2. Figure 1-6 on the facing page is drawn from NOAA Chart 18746, San Pedro Channel, at Newport Beach, California. It has been reduced from 1:80,000 to 1:160,000. Contour the soundings using a 50-fathom contour interval and then draw in the 10-fathom contour.

(a) Compare the scale of this chart with that of the Gulf of Mexico chart. Which chart depicts the greater area? _____

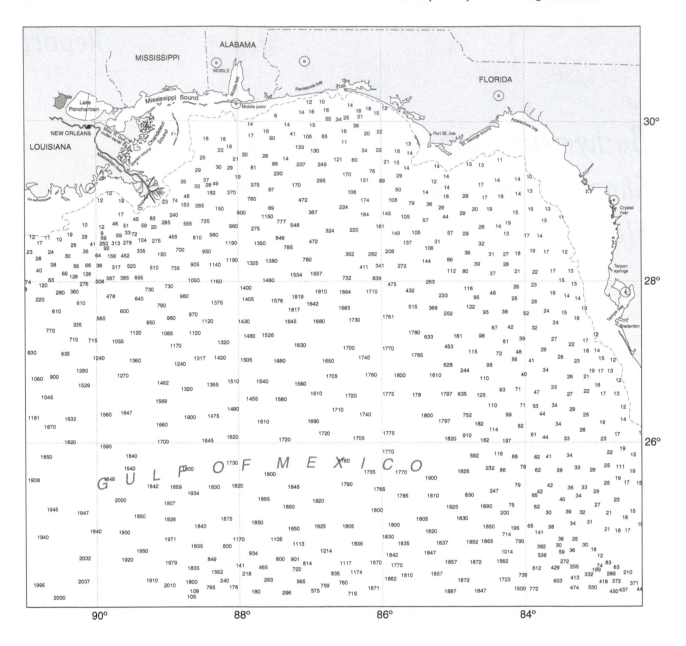

Figure 1-5 NOAA Chart 411, Gulf of Mexico.

Which chart yields the most detail about the sea floor? _____ Give a brief reason for your answer. _____

_____

_____

_____

_____

(b) On the graph paper of Figure 1-7 construct cross sections along the two profiles AB and XYZ shown on the chart. The upper graph is vertically exaggerated and the lower one has the same horizontal and vertical scales. The results illustrate why we exaggerate the vertical in marine work. Calculate the vertical exaggeration and write it in the space provided below the grid.

(c) Draw a heavy dashed line along the axis of the Newport Submarine Canyon. How deep is it along the section line Y-Z? _____ fathoms _____ meters.

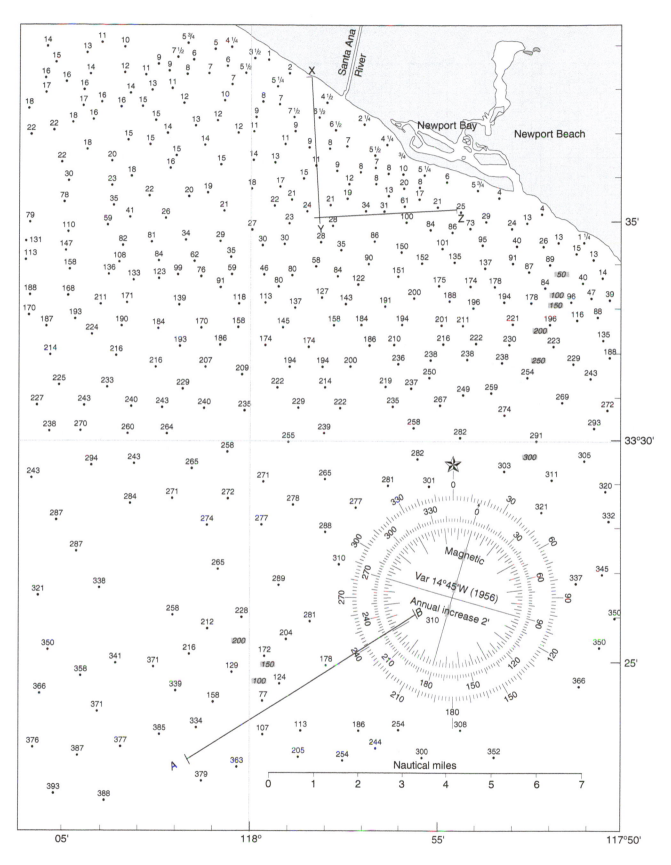

Figure 1-6   NOAA Chart 18746, San Pedro Channel, Newport Beach, California.

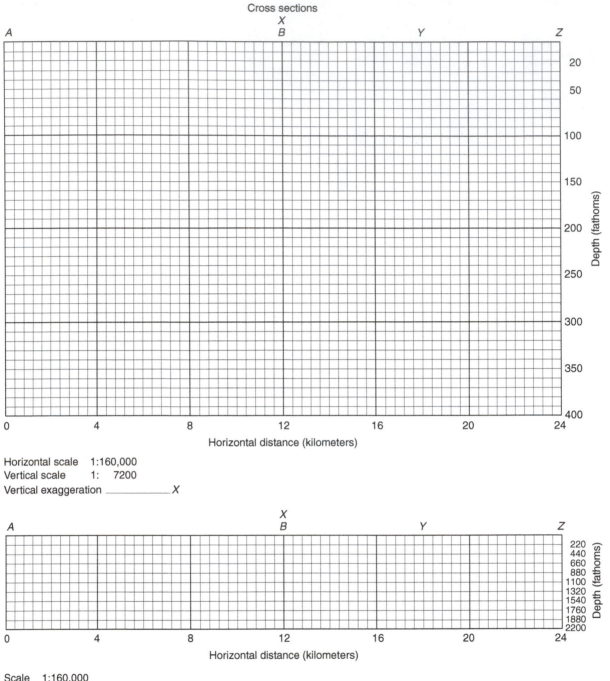

Cross sections

Horizontal scale   1:160,000
Vertical scale   1:   7200
Vertical exaggeration _____ X

Horizontal distance (kilometers)

Depth (fathoms)

Scale   1:160,000
No vertical exaggeration

Figure 1-7

3. Table 1-1 contains water depths and the distance from Jacksonville, Florida, to Agadir, Morocco, along the 31° north parallel. Draw a cross section of the Atlantic Ocean floor along this line on the graph provided in Figure 1-8. What is the vertical exaggeration of this cross section and why is it so large? _____

TABLE 1-1

Bathymetry of the North Atlantic Ocean at 31°N

Topographic Feature	Distance (nautical miles)	Depth (fathoms)
Florida shoreline	0	0
Florida Continental Shelf	95	100
	250	1000
	315	2000
	375	2450
Blake Outer Ridge	565	2200
Hatteras Abyssal Plain	755	2850
Bermuda Rise	1040	2700
North American Basin	1130	3000
	2300	3000
Mid-Atlantic Ridge	2610	2000
	2765	1810
	3020	2000
Cape Verde Abyssal Plain	3835	3000
	3900	3440
	3960	3000
	4465	2000
Moroccan Continental Shelf	4840	1000
	4900	100
Moroccan shoreline	4935	0

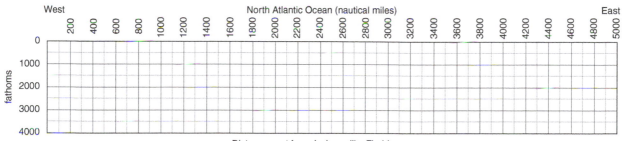

Figure 1-8

4. See the Redondo Submarine Canyon depth recording (Figure 1-4) for the following questions.

   (a) What is the water depth at the canyon axis in fathoms _____ ? meters _____ ?

_____

_____

_____

   (b) Water depth (D) (fathoms or meters) is equal to the velocity of sound in seawater (V) (fathoms/second or meters/second) multiplied by the one-way travel time (T) (seconds):

$$D = V \times T.$$

At a sound velocity of 1450 meters/second (800 fathoms/second), how long would it take a sound pulse from a ship directly over a canyon 3000 m deep to go from the ship to the bottom and return? _____ seconds. Show your work.

*M*arine Charts — or Finding Your Way Around

OBJECTIVES:

■ To understand the coordinate system (latitude and longitude) of the earth's sphere.

■ To become familiar with the multitude of data that are to be found on marine charts.

■ To learn to plot a pre-determined course on a chart and to navigate using directions and landmarks.

Since oceanographers rely on surface vessels for transportation, they use marine charts as base maps upon which they can locate their position and plot their data. For this reason it is worth our while to learn some of the basic principles of navigation and seamanship. Navigation has evolved from an art into a science in the course of its 6000-year history. Now, as in the past, the basic tools are the chart, the compass, and a method of determining position. The most critical requirement of all marine work may be precise positioning, because data reported from generalized positions are virtually useless to others wishing to follow up on previous work. This precision would be particularly important in the mining of hard minerals off the sea floor or in offshore oil exploration.

The Coordinates

The coordinates of **latitude** and **longitude** are essential in navigation. The lines of latitude are also called parallels of latitude, because they are parallel to the equator and to each other. Measured in degrees of arc along a circle, they specify the angular

distance north or south of the equator, from 0° at the equator to 90° at either pole. Each degree is divided into 60 minutes of arc (1° = 60′) and each minute into 60 seconds (1′ = 60″). Latitude is recorded with its hemisphere notation, north or south; for example, that of Seattle, Washington, is 47°36′N (Figure 2-1). Lines of longitude, or meridians, are also expressed in degrees and refer to the angular distance on the earth measured from the prime meridian (0°) at Greenwich, England, east or west through 180°. Longitude too should be reported with its hemisphere notations, east or west: for example, Seattle, 122°20′W.

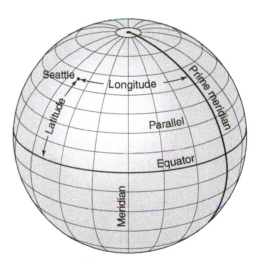

Figure 2-1 Meridian and parallel coordinate system for the earth. The prime meridian is 0° longitude, and the equator is 0° latitude. [After Nathaniel Bowditch, *American Practical Navigator*, Hydrographic Office Publication No. 9, U.S. Naval Oceanographic Office, 1966.]

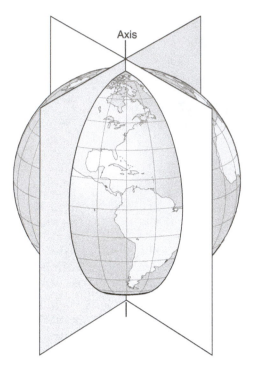

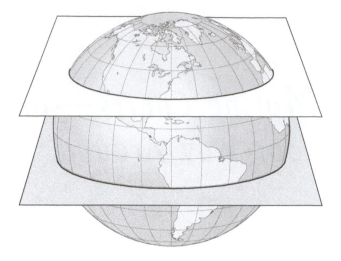

Figure 2-3 Parallels of latitude are parallel to the equator and are small circles. The equator is a great circle perpendicular to the meridional great circles. [After Nathaniel Bowditch, *American Practical Navigator*, Hydrographic Office Publication No. 9, U.S. Naval Oceanographic Office, 1966.]

Figure 2-2 The planes of meridians (lines of longitude) meet at the polar axis. [After Nathaniel Bowditch, *American Practical Navigator*, Hydrographic Office Publication No. 9, U.S. Naval Oceanographic Office, 1966.]

Another term that is frequently used in navigation is **great circle.** It refers to any circle traced on the surface of a sphere by a plane that passes through the center of that sphere. All longitudinal lines are great circles (Figure 2-2). **Small circles** refer to the lines of intersection of a sphere and a plane that do *not* pass through the center of the sphere. Lines of latitude are small circles except for the equator, which is a great circle (Figure 2-3).

The Marine Chart

Oceanographers and sailors frequently use charts of different scales, depending upon their proximity to land. For example, offshore navigation in the open ocean requires a large-scale chart depicting large areas of an ocean basin with little shoreline detail. Figure 1-5 is an example of such a chart. In contrast, nearshore navigation requires small-scale charts with detailed information (see Figure 2-6).

Most marine charts give water depths, the configuration of the shoreline in coastal waters, and other navigation aids such as lights and important landmarks. The title block is prominently placed on the chart and contains general information. Here will be found the name identifying the waters covered by the chart, the units of depth (feet, fathoms, or meters), and the datum plane, or sea-level reference, for the depth measurements, or soundings (Figure 2-4). In most charts, north is the top of the sheet, latitude scales are given on the sides, and longitude scales are at the top and bottom. Meridians and parallels are drawn at given intervals in fine black lines across the chart. The nature of the bottom is specified as hard (hrd), rocky (rky), gravel (g), sand (s), coarse (crs), shells (sh), or coral (co).

The coordinates latitude and longitude are used to locate, or "fix," a vessel's position on a marine chart. To do so, first find the desired latitude on the scales at either side of the chart and connect the two points. Now locate the desired longitude at the top and bottom of the chart and connect them. Where the lines cross will be the designated position. Note that many charts cover an area of less than 1° of latitude or longitude, so that the scales on the sides of the charts will be in minutes or seconds rather than degrees—an important point to remember when plotting positions.

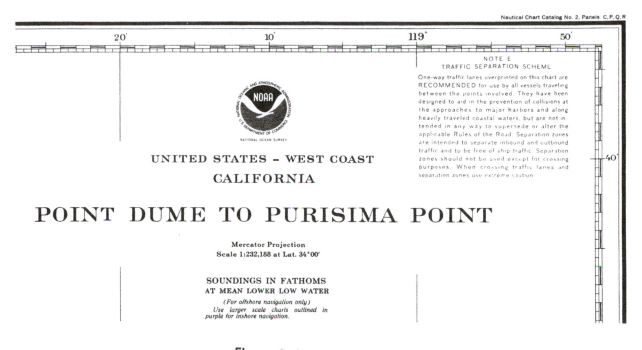

Figure 2-4 A chart title block.

Units of Distance and Speed

On land, distances are expressed in kilometers or in **statute miles,** whereas at sea they are in **nautical miles.** The nautical mile is equal to 1′ of latitude. A nautical mile is about 1.15 statute miles. Thus 1° of latitude is 60 nautical miles, or 60 × 1.15 = 69 statute miles. Most charts have a bar scale that shows distance in nautical miles or yards. To determine the distance of a given length on the chart, take an ordinary drafting compass or a pair of dividers and spread the points so that they touch the ends of the length to be converted. Transfer the compass points directly to the bar scale to obtain the distance.

The unit of speed used at sea is the **knot** and is defined as 1 nautical mile per hour. It is incorrect to speak of knots per hour because this means nautical miles per hour per hour, a unit of acceleration rather than speed. To convert knots to kilometers per hour, multiply by 1.85.

Plotting a Course

In plotting a course at sea we must distinguish between *magnetic* north (to which the north-seeking pole of a magnetic compass points) and *true,* or geographic, north. Geographic north is the North Pole, whereas a magnetic compass is attracted to a pole some distance from the geographic pole. The difference in degrees is known as **variation** (14°45′W in Figure 2-5). A *compass rose* appears on all navigational marine charts and shows the magnetic and true north directions. Almost all modern marine compasses are graduated from 0° (north) clockwise through 360°. There are 32 points on the compass, the four cardinal points being north (0°), east (90°), south (180°), and west (270°). The four intercardinal points, midway between the cardinal points, are northeast, southeast, southwest, and northwest. The points between each cardinal and intercardinal point, eight in all, are named for the directions they fall between: for example, north northeast (lying between north and northeast, or at 22.5°), east northeast (67.5°), south southeast (157.5°), and so on (Figure 2-5). The recitation of the full compass circle, known as "boxing the compass," was a remarkable accomplishment by early mariners, but a necessary one. Directions are now expressed largely by degrees rather than by points on a compass, except that cardinal and intercardinal points are used to indicate general directions, as in "northeast wind."

A ship's **course,** expressed in degrees, is the *intended* direction of travel: for example, a course of

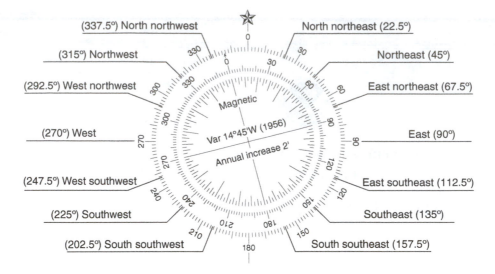

Figure 2-5 The compass rose.

180° is due south, and one of 135° is southeast. However, winds, ocean currents, and pilot error may prevent the ship from adhering to a particular course. A ship's **heading** or track is the direction in which the ship is actually traveling regardless of its prescribed course. Many times, the ship's heading is also called the **course made good,** to indicate the "true" course of the ship under the influence of winds and currents, which may deflect the ship from its intended course. All courses and headings are established in reference to true north unless otherwise indicated. A **bearing** is the direction from one point to another and is expressed as an angle from north.

Electronic Navigation Aids

Navigation at sea may be assisted by using a **LORAN-C** or **GPS** receiver. LORAN-C is a long-range navigation system using very low frequency (100kHz) pulsed radio signals broadcast from transmitters on shore, which have useful ranges of 800–1200 nautical miles from the transmitters. Positions at sea may be calculated from the distance to two different LORAN-C transmitters. Marine charts from the Defense Mapping Agency Hydrographic/Topographic Center are printed with LORAN-C "lattices," grids of lines needed for navigation with a LORAN-C receiver.

Another type of radio-navigation is **GPS**—the Global Positioning System. GPS receivers may be used to locate one's position, either on land or at sea, by triangulation. GPS receivers locate the distance to orbiting satellites by measuring how long it takes a radio signal broadcast from the satellite to reach the receiver. There are twenty-four GPS satellites orbiting the Earth, so four to eight satellites will be above the horizon from any point on Earth's surface. GPS receivers allow one to locate position to plus or minus a few meters' accuracy.

Time and the Earth's Rotation

Because the earth rotates once on its axis in the course of 24 hours, and a "day" thus is 24 hours long, it can also be said that one 360° rotation is equal to 24 hours. When the sun is directly over Greenwich, England (longitude 0°), the local time is noon, whereas at longitude 180° the local time is midnight. Thus every 15° of longitude east or west of Greenwich time equals 1 hour (coordinated universal time, sometimes called *Zulu,* or *Z, time*); or 360°/24 = 15° longitude per hour.

DEFINITIONS

Bearing. The horizontal angle between a line connecting an observer and the point being viewed, and a reference direction, usually north. It is recorded as an angle from north (000°) clockwise through 360°.

Course. The direction in which a ship must travel to arrive at a desired destination.

Great circle. The circle traced on the surface of a sphere by any plane that passes through the center of the sphere. All longitudinal lines are great circles.

Heading. The direction of actual travel, regardless of the prescribed course. The heading may be right on course or it may deviate as a result of various influences.

Knot. The unit of speed used at sea. It is equivalent to 1 nautical mile per hour.

Latitude. Angular distance north or south of the equator measured from 0° at the equator to 90° at the poles.

Longitude. Angular distance on the earth measured from the prime meridian (0°) at Greenwich, England, east or west through 180°.

Nautical mile. The basic unit of distance at sea, equivalent to 6080 feet, 1853 meters, or 1.853 kilometers.

Small circle. Any line of intersection of a sphere and a plane that does not pass through the center of the sphere. Lines of latitude are small circles parallel to the equator.

Statute mile. A unit of distance used on land, equivalent to 5280 feet, 1609 meters, or 1.609 kilometers.

Exercise 2

Marine Charts

NAME _____

DATE _____

INSTRUCTOR _____

1. Answer the questions using the following table:

City	Latitude	Longitude
Seattle, Washington	47°36′N	122°20′W
Honolulu, Hawaii	21°18′N	157°50′W
New York, New York	40°30′N	73°58′W
Boston, Massachusetts	42°15′N	71°07′W
Miami, Florida	25°45′N	80°11′W
Galway, Ireland	53°16′N	9°05′W
Bergen, Norway	60°24′N	5°20′E

(a) Which city is farthest north? _____

(b) What is the time difference to the nearest minute between New York and Bergen? Note that Bergen is east of the prime meridian. Convert minutes to decimal fractions of degrees _____ hours _____ minutes.

(c) What is the time difference between Seattle and Honolulu to the nearest hour ($=$ time zone)? _____ hours.

(d) How far north is New York from Miami? _____ nautical miles _____ kilometers.

(e) Which city in the United States is farthest east? _____

2. A speedboat moves at 35 knots (nautical miles per hour), and an auto moves at 35 statute miles per hour. Convert the speedboat's velocity to statute miles per hour and the auto's velocity to knots. Auto _____ knots; speedboat _____ miles per hour. Show your work.

Which is faster? _____

3. The following questions relate to NOAA Chart 11373, Mississippi Sound and Approaches, which has been reduced 70% in Figure 2-6 from its original 1 : 80,000 scale. Soundings on this chart are in feet. The questions require parallel rulers or two identical 30° – 60° triangles and a drafting compass.

(a) Your ship is located at 30°18′30″N, 88°57′30″W. Carefully locate this position on the chart, and mark it with a dot. Double-check your location (this is standard practice for marine navigation), then label this dot with numeral 1. What depth is the water at this station? _____

(b) From Station 1, determine the bearing to the Flag Tower on Biloxi Beach. The compass rose on the chart is an aid for measuring direction. Place a straightedge on the two points and transfer that direction to the rose by parallel motion with parallel rulers or two triangles.
What is the bearing to the Flag Tower in degrees true? _____. In degrees magnetic? _____.

(c) Determine the bearing from Station 1 to the Intracoastal Waterway Buoy (labeled Buoy A on the chart).
    What is the bearing to Buoy A in degrees true? _____. In degrees magnetic? _____.

(d) From Station 1, you set off on a course for the tower at Ship Island, when your ship sails into a fog bank. Using radar, the navigator determines you are 3.5 nautical miles away from Buoy A and 3.0 nautical miles from the tower at Biloxi. With a compass, locate your position off the harbor entrance and determine the actual course made good and the ship's distance from Ship Island.
    Course made good: _____ degrees true, _____ degrees magnetic. Distance: _____

(e) Your ship's draft is 7 feet. Do you see a problem if you continue on your present course? Explain.
_____
_____
_____
_____
_____
_____
_____
_____
_____
_____

(f) You decide to change your ship's heading to sail west of Ship Island by the Gulfport Ship Channel. To enter the Ship Channel at Buoy 12, to what course will you need to change?
    New course is: _____ degrees true, _____ degrees magnetic.
    Distance to Buoy 12 is: _____ nautical miles.

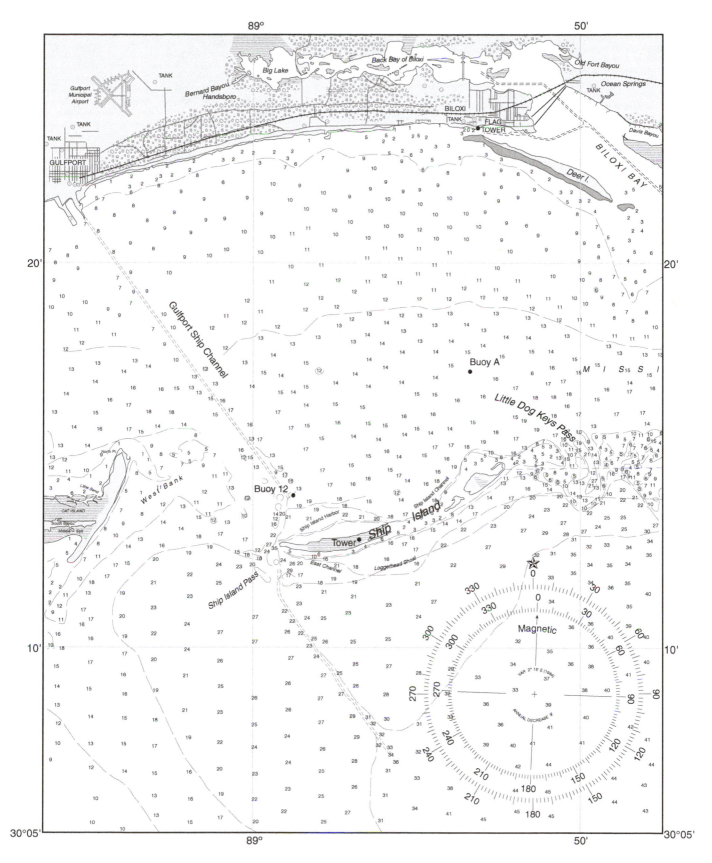

***Figure* 2-6** A portion of NOAA Chart 11373, Mississippi Sound and Approaches—
Dauphin Island to Cat Island. This chart has been reduced 70 percent from its original
1:80,000 scale. Note: 10 minutes of latitude equals 10 nautical miles.

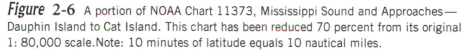

Sea–Floor Spreading and Plate Tectonics

OBJECTIVES:

■ To understand the distribution of the earth's major physiographic features such as volcanoes, oceanic trenches, and mid-ocean ridges.

■ To recognize the scientific importance of plate boundaries and the geophysical activities taking place there.

■ To investigate the slow rates at which plate tectonic processes take place and gain an appreciation of the great spans we call "geologic time."

It has long been recognized that such features as mountains, earthquakes, and volcanoes are not randomly distributed upon the surface of the earth. In 1912 the German meteorologist Alfred Wegener attempted to explain this distribution by arguing that if continents could move vertically, then horizontal motion, or *continental drift,* was also possible. The theory of *sea-floor spreading*—a process whereby new sea floor is created as adjacent crust is moved apart to make room—was first proposed in 1960. These two theories were then joined into a unifying model of crustal development called *plate tectonics.* Geophysical evidence collected over the course of the past three decades has given the theory of plate tectonics firm scientific support.

The Theory of Plate Tectonics

In essence, it has been proposed that the earth's outer shell (called the **lithosphere,** a zone about 100 kilometers thick) is composed of a number of rigid plates. Each of these **lithospheric plates** moves in a different direction, and thus plate boundaries are sites of tectonic activity where earthquakes, volcanism, and mountain building occur. Seven large plates and a number of smaller ones have been identified (Figure 3-1). The source of energy for driving plate motions is the earth's internal heat and gravity. The earth's interior is quite hot and provides a constant flow of heat to the earth's surface. Because of this, thermal gradients develop in the earth's interior, causing hot, low-density material to rise toward the surface while cold, higher-density material sinks. The convective motion thus generated in the earth's interior moves the lithospheric plates around on the surface of the earth, aided by gravity.

There are three types of plate boundaries: divergent boundaries, convergent boundaries, and transform faults (Figure 3-2). Where plates are diverging, **magma** rises and cools to form new ocean crust. This type of boundary is called a divergent boundary or **spreading center** and is expressed topographically as a oceanic ridge. Examples of spreading centers include the Mid-Atlantic Ridge and East Pacific Rise. Of course, if new sea floor is continuously created at spreading centers, old sea floor must be destroyed. This occurs at convergent boundaries, also known as **subduction zones,** which are expressed topographically by the presence of trenches. As old oceanic crust is subducted, it will be heated, causing partial melting. Magma produced in this fashion will rise, causing volcanism on the overriding plate near the trench. Examples of subduction zones include the Aleutian Trench and Aleutian Islands, and the Peru–Chile Trench and Andes Mountains. The

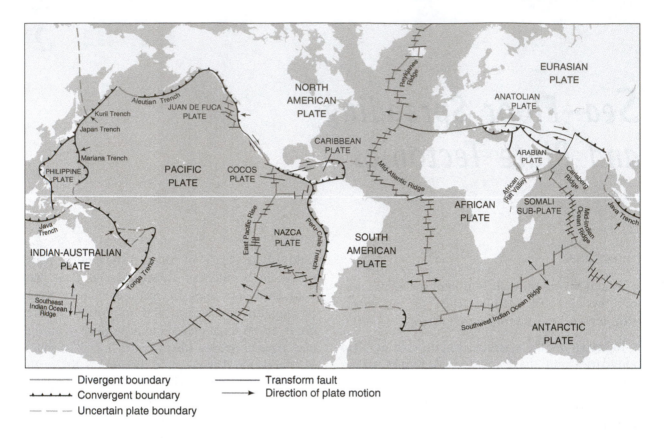

---- Divergent boundary ──── Transform fault
▲▲▲▲ Convergent boundary ───→ Direction of plate motion
- - - - Uncertain plate boundary

Figure 3-1 Major lithospheric plates and their boundaries. The plates are rigid lithosphere about 100 kilometers thick. They are in constant motion, and they interact with one another based upon their relative motion. At divergent boundaries, plates move apart and new crust is created. The spreading centers at mid-ocean ridges are divergent boundaries. At convergent boundaries, plates move together and one plate plunges under the other in the process of subduction. Thus, new crust is created at divergent boundaries and old crust is destroyed at convergent ones. Plates that slip horizontally past one another form the third type of boundary, known as a transform fault. [From Eldridge M. Moores, Editor, *Shaping the Earth: Tectonics of Continents and Oceans;* Readings from Scientific American, W. H. Freeman and Company, New York]

third type of boundary occurs where plates slide past one another with no crustal creation or destruction (Figure 3-2). These are also known as **transform faults** and may be expressed topographically on the sea floor as *fracture zones*. Examples of these are the Romanche, Oceanographer, and Mendocino fracture zones. The San Andreas Fault in California is the boundary between the Pacific Plate and the North American Plate.

Plate boundaries are seldom smooth, and thus plates "catch" on each other, deforming elastically on the edges until rock failure occurs. Failure produces the abrupt motion we know as earthquakes. The largest and most devastating of these events often occur in subduction zones, but transform

fault boundaries such as the San Andreas (in California) may also produce very large earthquakes. Occasionally two continental plates will collide in a subduction zone, but because continental crust has too low a density to sink into the earth's interior, the result is the pushing up of large mountain ranges. Collisions such as these created the Himalayas and the Alps.

Paleomagnetism as a Recorder of Sea-Floor Spreading

The critical evidence for sea-floor spreading is based on earth magnetism (the force exerted by the

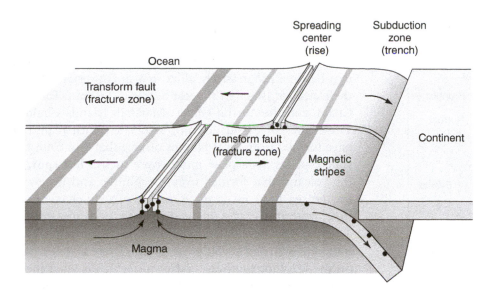

Figure 3-2 The major features of plate tectonics. New sea floor is created at a spreading center, or rise, as magma rises from the mantle below. The magma solidifies as lava in cracks and flows, thereby preserving the magnetic polarity of the earth's magnetic field at the time the cooling took place. A subduction zone, or trench, forms where an oceanic plate collides with a continental plate. A transform fault forms where two parts of a plate slip past one another, producing earthquakes (the foci, or points of rupture, are indicated by the heavy dots). Note that shallow-focus earthquakes (occurring close to the earth's surface) take place along transform faults and under rises, and that deep-focus earthquakes occur at subduction zones where one plate plunges beneath another. [After Don L. Anderson, "The San Andreas Fault." Copyright © 1971 by Scientific American, Inc. All rights reserved.]

earth's magnetic field). In the past (approximately every half million years), the earth's magnetic field has reversed its polarity, so that the north magnetic pole becomes the south magnetic pole (or the present-day *normal* polarity becomes *reverse* polarity). Each major reversal is termed a *polarity epoch.* The sequence of reversals occurring in the course of the past several million years has been dated with the use of radiometric techniques (Figure 3-3). Shorter reversals, lasting from tens to hundreds of thousand years, are termed *events,* and have also been recorded within the longer epochs.

When basalt is intruded and cools in cracks at mid-ocean ridges, the polarity of the earth's magnetic field at the time of cooling is preserved. As the crust moves away from the spreading center, each successive unit of the cooled magma gradually moves outward, revealing zones of *paleomagnetism* (Figure 3-4), in which the orientation of the earth's magnetic field at the time of formation is preserved. This preservation of paleomagnetism provides a method for measuring the rate at which new sea floor is formed.

Where the rocks have the same magnetic polarity as the present-day field, we find stronger than average magnetic field and we have a positive anomaly; where the rocks preserve reverse polarity, we measure weaker than average magnetic field and we have a negative anomaly. We can determine the rate at which new sea floor is formed by measuring the distance from the ridge crest to a magnetic anomaly of known age. To calculate the spreading rate, simply divide the distance traveled by the age of the oldest reliably dated anomaly. With the aid of this technique, plate velocities of 2–3 centimeters/year for the Atlantic Ocean and 16 centimeters/year for the Pacific Ocean have been calculated (Figure 3-5).

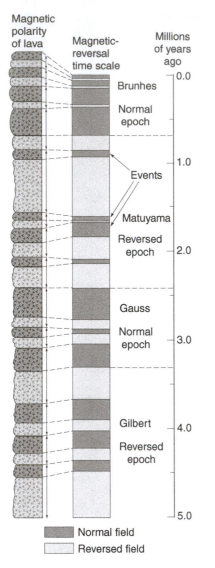

Normal field
Reversed field

Figure 3-3 Schematic illustration of how magnetic polarities of lava flows are used to construct the time scales of magnetic reversals over the past 5 million years. In no one place is the entire sequence found; the sequence is worked out by patching together the ages and polarities from lava beds all over the world. Note that each magnetic epoch is named after a famous magnetician. [From F. Press and R. Siever, *Earth*, 4th ed. W. H. Freeman and Company. Copyright © 1986.]

Hot Spots or Plumes

Within the Pacific Ocean basin there are chains of aligned volcanoes that become progressively older in a northwestern direction. They are the Hawaiian Islands, Emperor Seamount Chain, Tuamotu-Line Island Chain, and the Austral-Gilbert-Marshall Islands Chain (Figure 3-6a). These lines of islands are not associated with plate boundaries, but have formed from great plumes of lava that rise from below the lithosphere plates. As a plate moves over such a **hot spot,** it carries away the volcano that formed over the plume, and a new volcano forms on the sea floor in its place. A line of extinct volcanoes forms over time, with their ages increasing with distance from the hot spot. The island of Hawaii is the youngest of the Hawaiian Islands and is the only one with active volcanism today. There is a new volcano forming on the sea floor southeast of Hawaii, which is what theory would predict. A line drawn from 5-million-year-old Hawaii to 45-million-year-old Midway Island represents an arrow pointing in the direction of plate movement over the hot spot. The track makes a "lazy L" with the Emperor Seamount Chain, which extends in a more northerly direction from Midway. That chain traces a different direction of plate

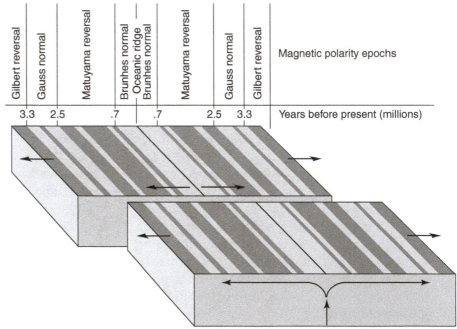

Figure 3-4 Evidence for sea-floor spreading has been obtained by determining the polarity of paleomagnetism in rocks lying on both sides of oceanic ridges. In the diagram, rocks of normal, or present-day, polarity are represented by the lighter gray stripes; rocks of reversed polarity are the darker gray. The displacement of the two blocks represents a fracture zone. The symmetry suggests that the rocks welled up in a molten or semimolten state and gradually moved outward. [After Patrick M. Hurley, "The Confirmation of Continental Drift." Copyright © 1968 by Scientific American, Inc. All rights reserved.]

movement prior to 45 million years ago. At least 122 hot spots have been active in the past 10 million years, and they are found on all the major plates of the earth, both oceanic and continental ones (Figure 3-6b).

Exotic Terranes and Plate Motion

In geology and geography the word "terrain" means the shape of the land, and the word "terrane" refers to the geology of the land. More specifically, "tectonostratigraphic terrane" or **exotic terrane** designates a fault-bounded block of rock with a history quite different from that of adjacent rocks

or terranes. At plate boundaries we find tearing, rifting, and collision of plates, so it is not surprising that bits and pieces of continental crust have torn away and traveled long distances to become part of a distant continent. Exotic terranes come in all sizes. India, for instance, can be viewed as a large terrane that broke away from an ancient southern landmass and traveled northward to collide with and become annealed to Asia. The terranes of western North America are elongated bodies or slivers of rock that moved along faults for thousands of miles to build coastal Alaska, British Columbia, and parts of California (Figure 3-7). In this manner continents grow and become

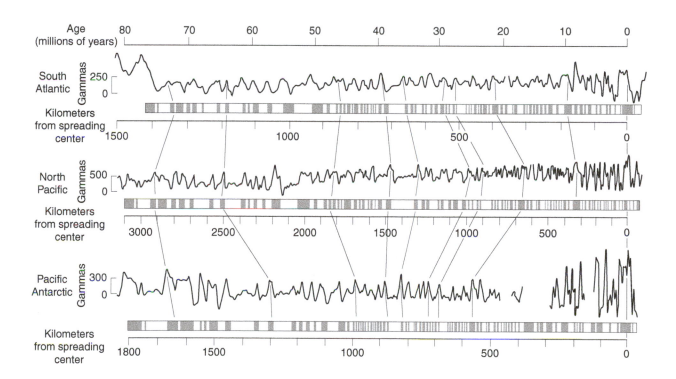

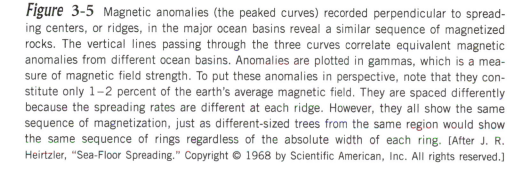

Figure 3-5 Magnetic anomalies (the peaked curves) recorded perpendicular to spreading centers, or ridges, in the major ocean basins reveal a similar sequence of magnetized rocks. The vertical lines passing through the three curves correlate equivalent magnetic anomalies from different ocean basins. Anomalies are plotted in gammas, which is a measure of magnetic field strength. To put these anomalies in perspective, note that they constitute only 1–2 percent of the earth's average magnetic field. They are spaced differently because the spreading rates are different at each ridge. However, they all show the same sequence of magnetization, just as different-sized trees from the same region would show the same sequence of rings regardless of the absolute width of each ring. [After J. R. Heirtzler, "Sea-Floor Spreading." Copyright © 1968 by Scientific American, Inc. All rights reserved.]

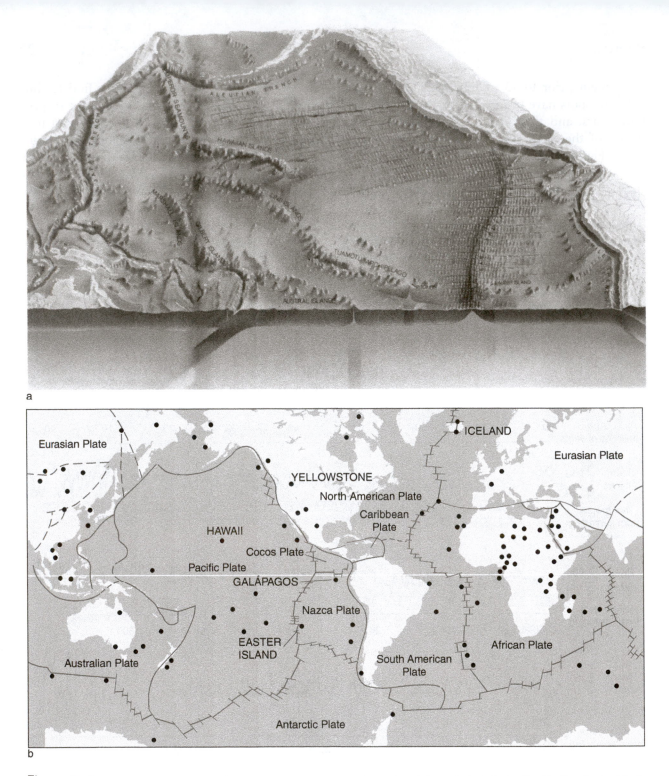

a

b

Figure 3-6 (a) Motion of the Pacific Plate over three hot spots has produced three parallel volcanic island chains. The chains lie within the plate, proving they were formed by a mechanism different from the one that produced the volcanic island arcs of Japan and the Aleutians, which are associated with subduction of the Pacific Plate trenches. [After Gregory Vink, E. Jason Morgan, and Peter R. Vogt, "The Earth's Hot Spots." Copyright © 1985 by Scientific American, Inc. All rights reserved.] (b) The population of hot spots includes at least 122 that have been active in the past 10 million years. They are found on all the major plates and on both oceanic and continental crust, but their distribution is decidedly nonuniform. Of the 122 hot spots, 43 are on the African Plate. Together with other evidence, this abundance of hot spots suggests that the African Plate is stationary over the mantle. If the African Plate is adopted as a frame of reference, other areas that have many hot-spot volcanoes, such as Antarctica and Southeast Asia, are found to be moving only slowly; on fast-moving plates hot-spot volcanism is rare. [After K. Burke and J. T. Wilson, "Hot Spots on the Earth's Surface." Copyright © 1976 by Scientific American, Inc. All rights reserved.]

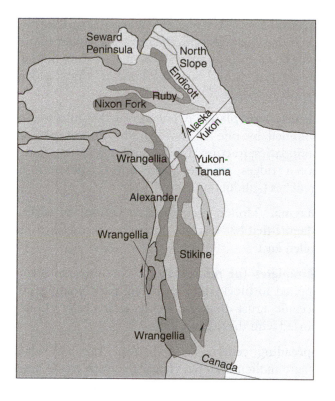

Figure 3-7 Accreted terranes of British Columbia and Alaska. The lightly shaded terranes (Wrangellia, Stikine, Ruby, and Nixon Fork) were once parts of other continents and have been displaced long distances. The stippled areas are probably displaced parts of North America, and the crosshatched areas are rocks that have not traveled long distances. The stable North American continent is darkly shaded.

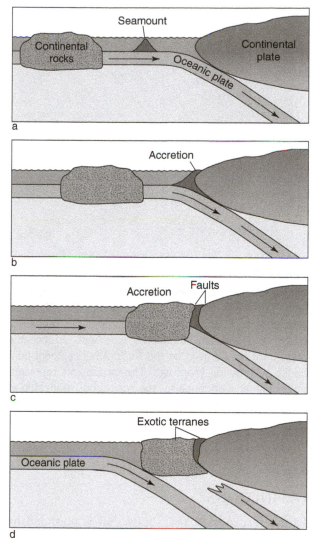

Figure 3-8 The formation of an exotic terrane by a microcontinent and a submarine volcano being scraped off a subducting plate onto the continental margin.

what amounts to a geologic collage. Terranes have paleomagnetic signatures that indicate a distant origin, and they have little in common geologically with adjacent terranes or the stable continental core.

Although terranes are found completely around the Pacific Ocean basin, western British Columbia and Alaska are the "prototypes" of exotic terrane discovery. Here we find a collection of **miniplates,** tectonic flotsam and jetsam if you wish, that have mashed into the North American continent over the past 180 million years (Figure 3-8). One block, Wrangellia, was an island during Triassic time, and it has a paleomagnetic signature indicating rocks that formed 16° from the equator. Whether Wrangellia formed north or south of the

equator is not known because we don't know if the earth's magnetic field was normal or reversed at that time. In either case the terrane traveled a long distance to become part of present day North America (Figure 3-9). It is estimated that 25 percent of the western edge of North America, from Alaska to Baja California, formed in this way; that is, by bits and pieces being grafted onto the core of the continent and thus enlarging it. Part of California west of the San Andreas Fault may become an exotic terrane of Alaska in the distant future.

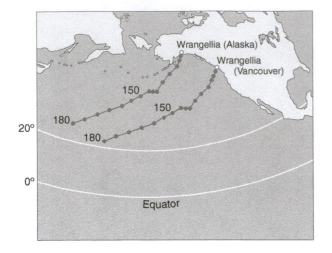

Figure **3-9** An analysis of Wrangellia terrane trajectories for the period between 180 and 100 million years ago, assuming an initial position in the Northern Hemisphere. In this interpretation, both the Vancouver Island and Alaskan miniplates rode with an oceanic plate until they arrived on the North American continent about 100 million years ago. The data points represent paleomagnetic evidence for the position of Wrangellia at 5-million-year time increments.

DEFINITIONS

Exotic terrane. A rock body that is bounded by faults and unrelated to rock bodies adjacent to it.

Hot spots. Fixed plumes of rising lava that have their origin in the mantle and are found within the ocean basins at the end of a chain of progressively older volcanoes or at divergent plate boundaries.

Lithosphere. The coherent rigid outer shell of the earth that includes both the crust and upper mantle. This layer is about 150 kilometers thick under the continents and 50–70 kilometers thick under the ocean basins.

Lithospheric plate. A portion of the lithosphere bounded by one or more of the three types of boundaries: fracture zones (transform faults); mid-ocean ridges or rises (spreading centers); and trenches (subduction zones).

Magma. Molten rock material within the earth. Magma that has reached the surface of the earth is called *lava*.

Miniplates (or microplates). An informal term applied to blocks of continental rock floating on oceanic crust that are being tectonically transported with the oceanic plate.

Spreading center. A mid-ocean rise or ridge where molten material (basalt) rises to create new sea floor.

Subduction zone. An elongate region along which a crustal plate descends below another one. For example, one such zone is the Peru–Chile Trench, along which the Nazca Plate descends beneath the South American Plate. During subduction, downward motion of the Nazca Plate produces earthquakes, and eventually the sinking rocks remelt to create local volcanic activity.

Transform fault. A special variety of lateral-slip fault along which the displacement suddenly stops. Where these faults offset mid-ocean ridges the actual slip is opposite the apparent displacement (Figure 3-2).

Exercise 3

Sea–Floor Spreading and Plate Tectonics

NAME _____

DATE _____

INSTRUCTOR _____

1. Name the type(s) of boundary (e.g., convergent, divergent, conservative) between the Pacific and North American plates at each of the following locations:

Southern California _____

Northern California, Oregon, and Washington _____

Aleutian Alaska _____

2. (a) What type of plate boundary dominates the circum-Pacific belt? _____

(b) What geologic hazards accompany this type of plate boundary? _____

_____

_____

3. Which of these stresses—*tension* (two forces acting in opposite directions), *compression* (forces acting toward each other in the same plane), or *shear* (forces acting toward each other in different planes)—characterize each type of plate boundary?

Divergent _____

Convergent _____

Transform fault _____

4. A mid-ocean ridge is offset along the transform fault in the sketch below. Place arrows to show the relative motion of the plates on opposite sides of the fault and indicate where one would expect earthquakes to be generated. Indicate where new sea floor is forming and the rock type.

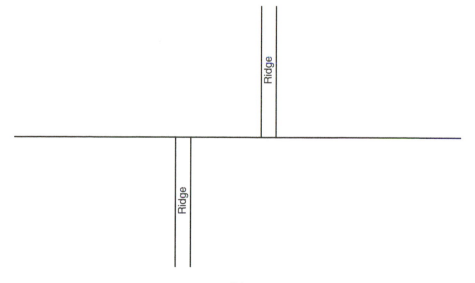

5. (a) For each magnetic anomaly record in Figure 3-5, determine the half-spreading rate in centimenters per year by first measuring the distance in kilometers to a known anomaly, then dividing this distance by the age of the anomaly, and finally converting kilometers/million years to centimeters/year. Show your work.

South Atlantic _____

North Pacific _____

Pacific Antarctic _____

(b) Which of the mid-ocean ridges has the fastest spreading rate? _____
   Which one has the slowest spreading rate? _____

6. In Figure 3-9 we see the trajectory of Wrangellia in its trek across the northwestern Pacific Ocean. Paleomagnetic data indicate it was near 20° latitude 180 million years ago and then "docked" (collided) in Alaska about 100 million years ago. Perform a "back of the envelope" calculation below to determine if Wrangellia's velocity is reasonable considering what we know about rates of plate motion. [Hint: Wrangellia moved across at least 40° of latitude. There are 60 nautical miles in a degree of latitude, so the terrane traveled some 2400 nautical miles, or 4440 kilometers, in 80 million years (there are 1.85 kilometers/nautical mile and 10^5 centimeters/kilometer—see Appendix A for conversion factors).] Simple multiplication and division will then tell us the annual rate at which Wrangellia has moved northward: _____ cm/year.

Is this rate reasonable in view of known rates of plate motion? _____
Assume that the eastward movement at these latitudes was about equal to the northward movement you calculated: _____ cm/year.
Is this reasonable in view of known rates of plate motion? _____

7. Find the following hot-spot lines of islands in Figure 3-6a: Hawaiian-Emperor Seamount Chain; Tuamotu-Line Island Chain; Austral-Gilbert-Marshall Islands Chain. Describe the change in direction of each of the hot-spot ridges or lines of islands, and then infer what the change implies about relative plate motion. _____
_____
_____
_____
_____

8. What would the earth be like if plate motion stopped; that is, how would it change physically, and what geophysical phenomena would increase, decrease, or cease altogether?_____
_____
_____
_____

Geography of the Oceans

OBJECTIVES:

■ To become aware of the diverse topography of the sea floor.

■ To arrange the major physiographic provinces in a systematic order.

■ To become familiar with the names, locations, and probable origins of many of the oceans' physical features.

This exercise was designed to be a "take home" exercise to help students become familiar with sea-floor geography and the names and locations of its major features. This knowledge should enhance your appreciation of other topics in oceanography, such as plate tectonics, ocean-current patterns, and the distribution of marine sediments and organisms.

Physiographic Relationships

Marine scientists have found that the largest features of the ocean basins have a regular and predictable pattern of interrelationships. For example, the principal submarine mountain chains, or ridges, traverse the major basins near the middle. They are flanked on each side by deep plains or plains with low hills. These abyssal plains and hills extend to the continents and terminate either in deep trenches or in gently sloping aprons of sediments. Such an apron, or blanket, of sediment is called the *continental rise* and consists of deposits that have accumulated at the base of the *continental slope*. Where an apron of sediment occurs at the base of a continental slope, it adjoins a deep flat plain of sediment. Where a trench occurs at the base of the slope, the neighboring deeper part of the ocean basin is hilly and has little sediment cover.

It has also been recognized that the mid-ocean ridges are broken by large fracture zones, or faults. These various features are shown in Figures 4-1 through 4-3. We will study them and their origins in more detail in this and other exercises.

Principal Morphologic Features

The major relief features of the ocean floor are listed in Table 4-1. The *first-order* morphologic features of the earth's crust are the continents and ocean basins. In the ocean basins, the *second-order* features consist of five types: **continental margins; deep-ocean floor; mid-ocean ridges** and **rises; fracture zones;** and **island arcs** and **trenches.** The first three types and their subdivisions are shown in Figure 4-1.

The five major second-order features and their smaller-scale, or third-order, features, as well as certain special forms, are described in the list of definitions at the end of this exercise.

TABLE 4-1

Major relief features of the ocean floor

First-order Features of the Earth's Crust

Continents
Ocean basins

Second-order Features of the Ocean-basin Floor

Continental margins
Deep-ocean floors
Mid-ocean ridges
Fracture zones
Island arcs and trenches

Third-order Features

Continental margins

Typical Atlantic forms

Continental shelf
Continental slope
Continental rise
Submarine canyons

Typical Pacific forms

Continental shelf
Continental slope
Marginal trough
Submarine canyons

Special forms

Continental borderland
Marginal plateaus

Deep-ocean floors

Abyssal plains
Abyssal hills
Abyssal fans
Seamount trends
Deep-ocean channels
Gaps and local rises
Atolls

Mid-ocean ridges

Crest provinces
Flank provinces
Rift zones

Fracture zones

Scarps
Depressions

Island arcs and trenches

Volcanic island arcs
Insular slopes
Trenches

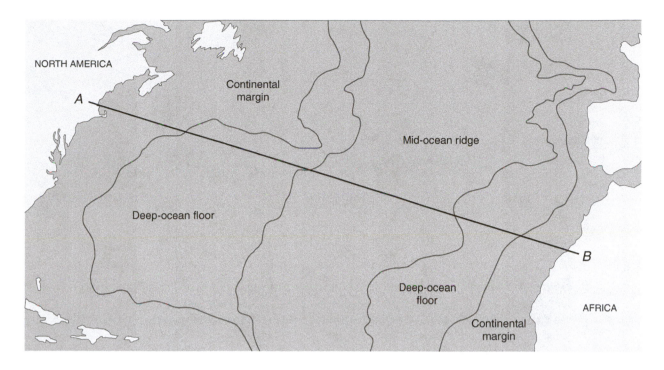

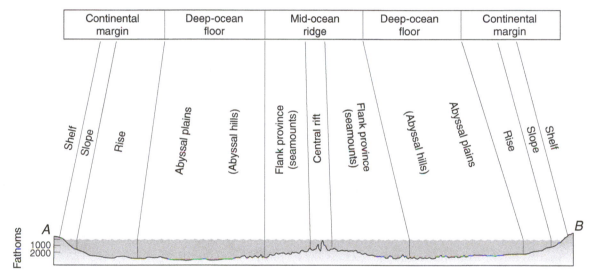

Figure 4-1 Some second-order morphologic features of the ocean basin. The top part is a simplified map of the Atlantic Ocean basin showing the continental margins, the deep-ocean floor, and the mid-ocean ridge. The bottom part is a profile of the sea floor along the line *AB,* showing the second- and third-order features of the North Atlantic Ocean basin. [After B. C. Heezen, M. Tharp, and M. Ewing, Geological Society of America Special Paper 65, 1959.]

Relationship Between Geographic Features and Major Processes

When we visualize the relationships among features of the ocean floor in charts or diagrams such as those shown in the figures, we are immediately led to ask questions. Why are the forms of the continental margins different from ocean to ocean? Why do some mid-ocean ridges occur in the centers of ocean basins? Why are rises, channels, and

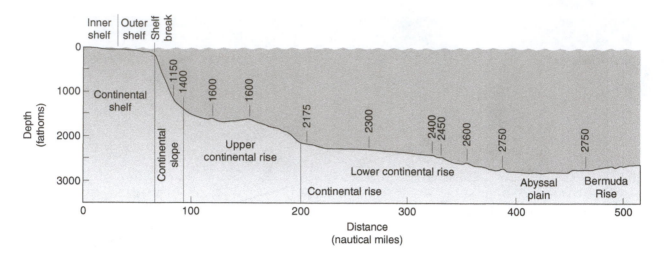

***Figure* 4-2** Profile of an Atlantic type of continental margin and its major morphologic subdivisions. The topography shown extends from the northeastern United States to the rise surrounding the island of Bermuda. [After B. C. Heezen, M. Tharp, and M. Ewing, Geological Society of America Special Paper 65, 1959.]

abyssal plains always associated? These association problems are answered for some of the major features when we discussed plate formation and motions in Exercise 3 on sea-floor spreading.

We can see in Figure 4-3 on pages 38–39 that margins around the Pacific Ocean are narrow and steep and commonly have trenches at the base of the slope. Margins around the Atlantic are wide and are typically bounded at the base of the slope by broad rises that merge with the deep abyssal plains. If we further recall that trenches are evidence of plate collisions and that wide continental margins with rises are moving on broad crustal plates, we can begin to understand the reasons for the differences in ocean-floor features in different oceans. We will discuss the processes that modify or form these features in other exercises.

DEFINITIONS

Active margin (between plates). A zone where ocean and continental plates collide. A deep trench is typically formed at the base of steep slopes. Shelves are narrow, and the shore is high and characteristically backed by a coast-parallel line of volcanoes. Earthquakes are common to great depths below the continental margin, and volcanic activity is common.

Continental margins. A zone consisting of continental shelf, slope, and rise. Several types of margins are recognized:

(1) The *Atlantic type* is characterized by a wide continental shelf, a steeper continental slope descending to the deep sea, and a flatter continental rise at the base of the slope formed by accumulation of sedimentary materials (Figure 4-2).

(2) The *Pacific type* is characterized by a narrow shelf and slope descending into a deep marginal trough, or trench, generally parallel to the continental margin. An example is the area of Chile and Peru in South America.

(3) The *marginal plateau* has a narrow shelf and incipient slope leading down to a shelflike feature similar to the continental shelf. The plateau is similar to the shelf, but occurs at a much greater depth. The Blake Plateau off Florida, at a depth of about 1000 meters, is a good example.

(4) The *borderland* consists of a series of offshore basins and ridges or islands, such as those found off southern California.

Deep-ocean floor. A zone characterized by abyssal plains of depositional (sedimentation) origin, abyssal hills, seamounts (volcanoes), guyots (flat-topped seamounts or tablemounts), channels, and gaps and local rises. (See Figures 4-1 and 4-2.)

Fracture zone. A linear zone of irregular topography on the sea floor, averaging 100 miles wide and more than 1000 miles long. These are, in fact, great faults along which sea-floor spreading has taken place, an example being the Eltanin Fracture Zone. Fracture zones are characterized by escarpments (scarps) several thousand feet high that separate regions of different depths. These zones are conspicuous for numerous seamounts, irregular topography, and local depressions. Unfortunately, some of these depressions have been called trenches, such as the Romanche Trench in the Atlantic, but they bear no relationship to deep-sea trenches in the Pacific Ocean.

Island arcs and trenches. Arc-shaped island chains generally curving seaward, with steep insular slopes and deep trenches on the seaward sides. They are found in the northern and eastern Pacific Ocean, and in the eastern Indian Ocean. Examples are Japan and the Japanese Trench, the Aleutians and the Aleutian Trench, and the Marianas Islands and Mariana Trench. In these trenches are the deepest places in the oceans.

Mid-ocean ridges and rises. Ridges are elongate, steep-sided elevations of the sea floor having a central rift (valley) and rough marginal topography that flanks the crest; an example is the Mid-Atlantic Ridge. Rises are broad, elongate, and smooth elevations of the sea floor, such as the East Pacific Rise. Both ridges and rises have a similar origin (see Exercise 3).

Passive margin (within plates). A continental margin with broad shelves, gentle wide slopes, and rises, typically found in locations where the margin of the continental block is part of a larger ocean–continental crustal plate, as in the Atlantic Ocean. Thus, the margin is "passively" moving on the plate. Such margins are generally areas of few earthquakes, no volcanic activity, and low heat flow.

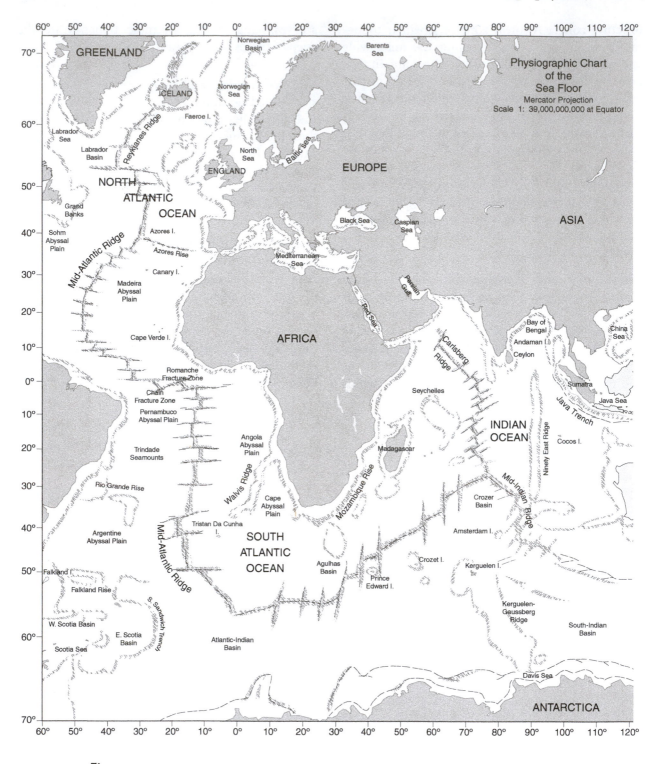

Figure 4-3 The ocean basins and their major features. [Courtesy Hubbard Scientific Company.]

120° 130° 140° 150° 160° 170° 180° 170° 160° 150° 140° 130° 120° 110° 100° 90° 80° 70° 60°

70°

Chukchi
Sea

Sea of
Okhotsk

Okhotsk
Basin

Aleutian
Basin

Bering
Sea

Hudson Bay

60°

NORTH AMERICA

50°

Emperor Seamount Chain

Aleutian Trench

Aleutian
Abyssal
Plain

Kuril Trench

Chinook
Fracture
Zone

Mendocino Fracture Zone

Sea of Japan

Japan Trench

Yellow
Sea

**NORTH PACIFIC
OCEAN**

Murray Fracture Zone

Baja California
Seamount Province

Bermuda
Rise

Hatteras
Abyssal
Plain

40°

30°

Bonin
Trench

Molokai Fracture Zone

Gulf of
Mexico

Nares Abyssal
Plain

Puerto Rico Trench

20°

Philippine
Sea

Mariana Trench

Mariana
Basin

Mid-Pacific Mountains

Clarion Fracture Zone

Caribbean Sea

10°

Philippine
Trench

Line I.

Clipperton Fracture Zone

Galápagos Fracture
Zone

Cocos Ridge

**SOUTH
AMERICA**

Solomon Rise

**NEW
GUINEA**

Vitas Trench

Samoa I.

Marquesas I.

Marquesas
Fracture Zone

Galápagos

Galápagos Rise

Pacific Rise

10°

Coral
Sea
Basin

North Fiji
Plateau

Fiji I.

Society I.

Nazca
Ridge

Peru-Chile Trench

Coral
Sea

Cook I.

Austral I.

Tuamotu
Archipelago

20°

Tonga I.

Kermadec-Tonga Trench

Easter Fracture Zone

Lord Howe Rise

AUSTRALIA

**SOUTH PACIFIC
OCEAN**

Easter

Juan Fernandez I.

30°

Great B. Reef

Australia Basin

Tasman
Basin

Tasman Sea

Southwest Pacific
Basin

Chile Rise

40°

Tasmania Rise

Chatham
Rise

Campbell
Plateau

Southeast Pacific
Basin

50°

Eltanin Fracture Zone

60°

Pacific-Antarctic Ridge

Balleny I.

Scott I.

70°

120° 130° 140° 150° 160° 170° 180° 170° 160° 150° 140° 130° 120° 110° 100° 90° 80° 70° 60°

Exercise 4

Geography of the Oceans

NAME _____

DATE _____

INSTRUCTOR _____

Report

1. On the map in Figure 4-3, locate the features or areas listed below. Refer also to the physiographic diagrams of the major oceans by B. C. Heezen and M. Tharp (published by *National Geographic*), if available. On Figure 4-4 below, place the number for the feature in the appropriate location.

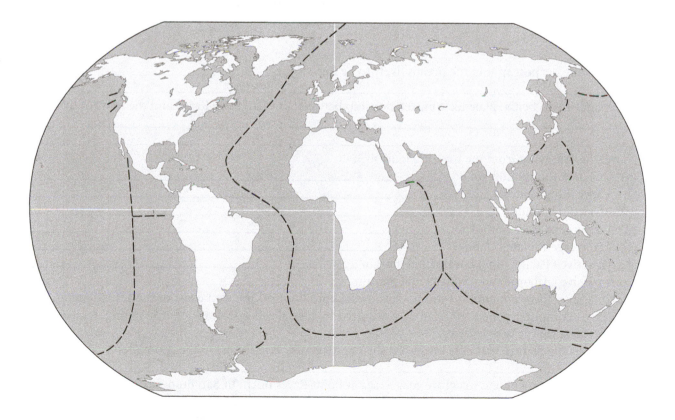

Figure 4-4 Worksheet for location of geographic features of the sea floor. The mid-ocean ridge system is indicated by the dashed lines.

Pacific Ocean	Atlantic Ocean	Indian Ocean
1. East Pacific Rise	1. Argentine Abyssal Plain	1. Carlsberg Ridge
2. Mendocino and Murray fracture zones	2. Mid-Atlantic Ridge	2. Seychelles Islands
3. Galápagos Islands	3. Bermuda Rise	3. Arabian Sea
4. Mariana Trench	4. Walvis Ridge	4. Kerguelen Islands
5. Peru–Chile Trench	5. Puerto Rico Trench	5. Madagascar
6. Sea of Okhotsk	6. Scotia Sea or Ridge	6. Red Sea
7. Bering Sea	7. Falkland Islands	7. Persian Gulf
8. Kermadec–Tonga Trench		
9. Emperor Seamount Chain		

2. What kind or class of fault is represented by the Romanche and Eltanin fracture zones? Make a rough sketch of the relative movement and location of earthquake epicenters on this type of fault (see Exercise 3).

3. Abyssal plains appear more extensive and continuous in the Atlantic Ocean than in the Pacific. Why might this be? _____
_____

4. What is the geographic relationship between the Mid-Atlantic Ridge, the Mid-Indian Ocean Ridge, and the East Pacific Rise?_____
_____

5. Make a freehand profile of a seamount (guyot) and explain how such features might have formed.

6. Where is the deepest spot in the oceans likely to be found? _____
_____

7. What appears to be the plate tectonic relationship between the Carlsberg Ridge and the Red Sea Trough?
_____
_____

8. Name the following from the map in Figure 4-3:
  (a) An active subduction zone _____
  (b) A spreading center _____
  (c) A transform fault _____
  (d) A divergent plate boundary _____
  (e) A convergent plate boundary _____
  (f) An active (Pacific type) continental margin _____
  (g) A passive (Atlantic type) continental margin _____

9. The image on the front cover of this book shows the sea floor off the west coast of the United States. The image is computer-generated from data acquired by specialized side-scan sonar mapping. Answer the following questions as best you can:
  (a) What is the feature that cuts across the shelf at Monterey Bay almost to the shoreline? _____
_____
  (b) What famous geologic structure goes to sea at Point Reyes north of San Francisco? (Hint: Note the trend of the linear valley on land south of San Francisco.) _____
  (c) Why is there no trench at the base of the shelf in this region even though it is a Pacific Ocean margin?_____
_____

Materials of the Sea Floor

OBJECTIVES:

■ To better understand the marine and terrestrial processes that determine the distribution of sediments and mineral resources on the sea floor.

■ To appreciate the role that shell-building organisms play in the accumulation of important sediment types and in moderating the chemistry of the oceans.

The variety of sedimentary materials on the sea floor is rich indeed. They consist of rock, mineral, and organic particles that have been transported by wind, water, or ice. A very small contribution comes from outer space in the form of tiny meteorites. Far from land, in the *pelagic* realm of the sea, much of the ocean floor is covered by the microscopic shells or "tests" of marine animals and plants that once lived in the water column. As they died, their tests sank to the ocean floor and formed organic **oozes** composed of calcium carbonate, opaline silica, or mixed calcareous-siliceous sediments (Figure 5-1).

Classification of Marine Sediments

Marine sediments may be classified into five broad groups according to the origin or source of the particles:

1. Biogenous deposits: Particles derived from biological processes. Shell matter, tests, and other remnants of life form these deposits. On the deep-sea floor they are called oozes.

2. Lithogenous deposits: Inorganic sediments composed of **rock** and **mineral** fragments of terrestrial origin. "Lithogenous" is used interchange-

ably with "terrigenous," but the latter usually refers to deposits on the continental shelves and slopes, whereas the former refers to pelagic sedimentation far from shore. Terrigenous deposits are classified by particle size, regardless of origin, from finest clays to very coarse sand and gravel (Table 5-1).

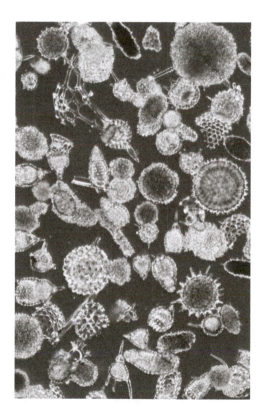

Figure 5-1 Light micrograph of "radiolarian ooze," a soft siliceous sediment that covers large areas of the deep ocean floor, consisting of the skeletal remains of radiolarians. Sample taken near Barbados. Rheinberg illumination; magnification 12.8x at 35mm. [M. I. Walker/Science Source. Enhancement by: Jessica Wilson.]

TABLE 5-1

Sediments and their grain size*

Name	Grain diameter (millimeters)	Grain diameter (micrometers)†
Gravel		
Boulder	> 256**	256×10^3
Cobble	$64-256$	$64-256 \times 10^3$
Pebble	$4-64$	$4-64 \times 10^3$
Granule	$2-4$	$2-4 \times 10^3$
Sand		
Very coarse	$1.0-2.0$	$1000-2000$
Coarse	$0.50-1.0$	$500-1000$
Medium	$0.25-0.50$	$250-500$
Fine	$0.125-0.25$	$125-250$
Very fine	$0.062-0.125$	$62-125$
Silt	**$0.004-0.062$**	**$4-62$**
Clay	**< 0.004****	**< 4***

*Size scale developed by C. K. Wentworth in 1922.
†One micrometer equals one one-thousandth (0.001) of a millimeter. It is the metric unit commonly used to measure fine-grained sediments and microscopic organisms.
**> means "greater than"; < means "less than."

Figure 5-2 Scanning electron micrograph of radiolaria. Magnification 800x @ (5x7inches). [Biophoto Associates/Science Source.]

3. Hydrogenous deposits: Sediments that form by chemical precipitation from seawater.

4. Volcanogenic deposits: Particles erupted from volcanoes. These particles are classified by size into volcanic ash (fine), lapilli, and blocks and bombs (very large).

5. Cosmogenous deposits: Primarily micrometeorites that did not burn completely while falling through the earth's atmosphere. They are rare and occur only in the deepest ocean basins far from land, where other sediment accumulates very slowly.

Biogenous Sediments

Biogenous deposits, known as oozes, consist of more than 30 percent skeletal debris. They may be divided into the taxonomic and chemical groups in the following list:

SILICEOUS OOZE($SiO_2 \cdot nH_2O$, opal)

Radiolarian ooze: Small silt-sized (50–100 micrometers) marine protozoa found in abundance in the equatorial waters of the oceans (Figure 5-2).

Diatom ooze: Small silt- to fine-sand-sized (10–500 micrometers) single-celled plants most abundant in polar regions with nutrient-rich waters (Figure 5-3).

CALCAREOUS OOZE ($CaCO_3$, limestone and chalk)

Foraminiferal ooze: Skeletons of sand-sized marine protozoans (1–300 micrometers). (Figure 5-4 top).

Coccolith ooze: Very tiny (clay-sized, 1–20 micrometers) plates and fragments of one-celled plants. (Figure 5-4 bottom).

Pteropod ooze: Small (1–2 millimeters) shells of a pelagic gastropod; covers a small area of the sea floor compared to other oozes.

Whether a biogenous ooze forms on the sea floor depends upon the amount of biological production in surface waters, its dilution by other sediments, and its destruction by scavengers or chemical solution.

Calcareous ooze is not known to form on the present sea floor below a depth of 4500 meters, even though there may be high production rates of calcareous organisms in the overlying waters. The reason is that below this depth, which is known as the "snow line," or **carbonate compensation depth** (CCD), the shells are dissolved.

The process and rate of carbonate dissolution are known only generally, but the top of the zone of complete dissolution of calcium carbonate ($CaCO_3$) usually coincides with the top of the **Antarctic bottom water** (see Exercise 8). This cold, high-pressure (deep) water mass forms in Antarctic surface waters as pack ice forms. It is saturated with carbon dioxide at the surface, but as it sinks toward the deep-ocean floor it becomes undersaturated because of the increased pressure. (Gas saturation is determined by temperature and pressure; solubility increases at low temperatures and at high pressures.) This increased dissolving capacity causes the carbonate particles that fall through this water to dissolve. Another factor related to carbonate dissolution is the amount of production of calcareous organisms in the surface waters. Where production is high, bacterial destruction of organic matter falling through the deeper waters uses up dissolved oxygen and increases the carbon dioxide content. These factors also increase the capacity of the deeper waters to dissolve carbonate. Therefore, increased production is compensated by increased solution, although the net result is carbonate addition to oceanic sediments on the sea floor. The CCD has

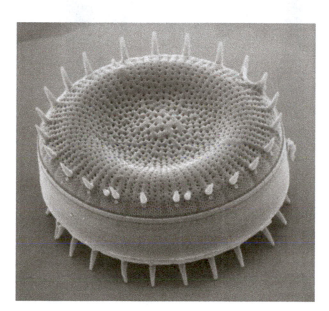

***Figure* 5-3** Scanning electron micrograph (SEM) of *Stephanodiscus astraea*, a species of centric diatom. The diatoms are a group of photosynthetic, single-celled algae containing about 10,000 species. They form an important part of the plankton at the base of the marine food chain. Diatoms are usually about 0.1 mm in diameter.
[BIOPHOTO ASSOCIATES/Science Source/Getty Images.]

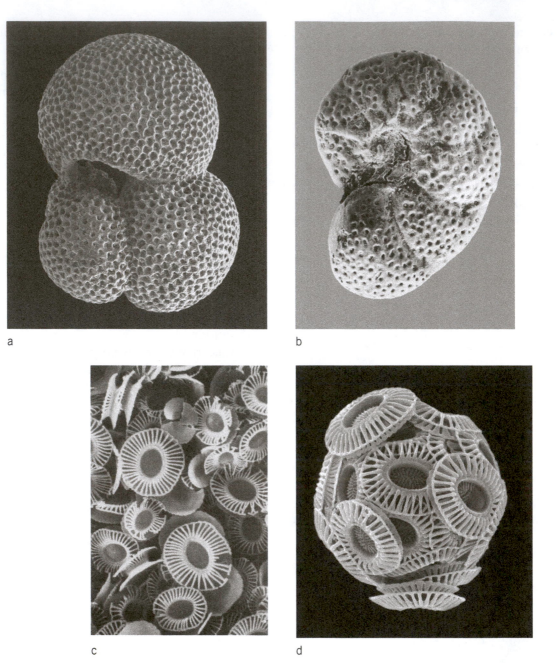

a

b

c

d

Figure 5-4 A) Colored scanning electron micrograph (SEM) of the shell of a foraminiferan, which are usually about 1mm in diameter. Magnification: ×140 when printed 10cm high. [DEE BREGER/SCIENCE PHOTO LIBRARY.]
B) Scanning electron micrograph (SEM) showing the fossilized foraminifera species Cibicidoides vulgaris, dating from the Paleocene epoch. Magnification: 260×, printed at 5×7" [Biophoto Associates/Science Source.]
C) Scanning electron micrograph (SEM) of fossilized coccoliths of the phytoplankton species Emiliania huxleyi taken from the White Cliffs of Dover. Coccoliths are components of the calcareous exoskeleton of nanoplankton. [Biophoto Associates/Science Source.]
D) Calcareous phytoplankton. Scanning electron micrograph (SEM) of Emiliana huxleyi. This small algal organism (coccolithophore) is surrounded by a skeleton (coccosphere) of calcium carbonate plates (coccoliths). When the organism dies, the plates separate and sink to the ocean floor. Individual plates have been found in vast numbers and can make up the major component of a particular rock, such as the Cretaceous-age chalk deposits of southern England. E. huxleyi is found in marine environments world wide. [Steve Gschmeissner/Science Source.]

varied over geologic time in relation to changes in ocean productivity. Its present level, at about 4500 meters, varies somewhat in each ocean basin.

Lithogenous Sediments

So-called brown mud or red clay covers about 38 percent of the deep-ocean floor and 28 percent of the total sea floor. Because of the great length of time it takes for this fine-grained material to settle through the water column, there is ample time for iron in the sediment or in the water to react with dissolved oxygen to create a brown or reddish coating on a sediment grain. These sediments are among the most slowly accumulating deposits known, the average rate being about 1 millimeter per 1000 years. The origin of these sediments is atmospheric dust, fine-grained rock debris from land, micrometeorites, volcanic ash and dust, and the insoluble residues from dissolved carbonate oozes. Red clay is found almost everywhere on the sea floor below 4500 meters, except for the deep trenches, where a mixture of coarser terrigenous material is to be found.

The term terrigenous, although interchangeable with lithogenous, is usually reserved for mostly land-derived (rock and mineral) deposits on or near the margins of the continents, regardless of origin. Thus a fine-grained deposit may be classified as silt even though it contains silt-sized shell fragments. The major source of terrigenous deposits in the oceans is a few large rivers. The three largest Asian rivers, the Ganges, Yellow, and Yangtze rivers, contribute about one-quarter of the world's total terrigenous sediment. The 20 largest rivers, almost all of which drain into the Atlantic Ocean, contribute approximately 75 percent. Because of the active plate margins and young mountain ranges that surround the Pacific Ocean, few large rivers enter directly into it.

Terrigenous deposits are difficult to characterize in a general statement because they accumulate in shallow or marginal areas subject to variability in supply and intensity of transporting agents. Deposits off mountainous coasts differ from those off coastal plains, and sediments derived from glacial terrains, such as Antarctica or Greenland, have distinct properties that indicate their origin. Thus each area of terrigenous sedimentation must be considered a unique environment, and only after intensive investigation can conclusions be made about the origin of the deposit.

Hydrogenous Sediments

Where the process of sedimentation is very slow, as in the red clay areas, manganese nodules may form (Figure 5-5). The nodules grow slowly, particle by particle, on a nucleus of skeletal material or mineral matter. Therefore, a slow rate of sedimentation of other components is required, or else they would be buried and removed from interaction with seawater. It is estimated that manganese nodules cover about 10 percent of the deep Pacific Ocean floor, and a somewhat smaller area of the Atlantic and Indian ocean basins. They are composed mostly of iron and manganese oxides, but

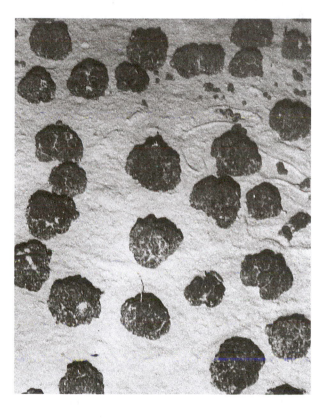

Figure 5-5 A field of rounded manganese nodules photographed on the abyssal sea floor at a depth of 5145 meters in the Pacific Ocean. Field of view of the photograph is about 0.5 meters from top to bottom, so these nodules are approximately the size of softballs. [From Bruce C. Heezen and Charles D. Hollister, *The Face of the Deep.* Copyright © 1971 by Oxford University Press, Inc. Reprinted by permission.]

are of economic interest for the small amounts of copper, nickel, and cobalt they contain.

Suspended Sediment Transportation

Much fine silt and clay remains in suspension after it has been introduced to the ocean from a river. The rate at which these small particles (usually less than 0.03 millimeter, or 30 micrometers, in diameter) settle is very slow, and so very slow currents can keep them suspended. However, coastal waters are also biologically very productive, and therefore the surface waters are populated by many microscopic and larger organisms that constantly filter or sweep the water for food particles. As they capture these particles they also capture these fine mineral particles, which pass through the gut and are excreted as aggregates of particles with some binding organic matter. Because they are usually large and dense, these fecal pellets fall rapidly. Thus the organisms "vacuum clean" the waters, stripping out particles and sending them to the bottom as pellets. This explains why almost 90 percent of the terrigenous sediments contributed to the oceans settle out on the margins within a few hundred kilometers of their sources. This also explains why so little fine sediment escapes to the deep-ocean floor, and why the pelagic clays are mainly wind-borne dust or oceanic volcanic dust.

Bottom-Transported Pelagic Sediments

In addition to particles that settle grain by grain on the deep-sea floor, there are more coarsely textured bottom-transported components that reach abyssal depths. These materials are transported by dense, sediment-laden turbidity currents that slide from continental slopes or submarine canyons and travel along the bottom to the deepest ocean floor. Such currents have actually been observed and are known to attain velocities greater than 20 kilometers per hour. Transport by turbidity currents explains sand and silt layers, some containing land plants and shallow-water shells, that are found in abyssal-plain and trench sediments. As the sediment load in the current settles out, distinctive deposits, called *turbidites,* in which individual grains grade upward from coarse to fine, are formed.

These currents suggest a transporting mechanism for part of the thick blanket of sediment found on the abyssal plains of the Atlantic Ocean and for the great wedge of terrigenous material trapped in deep trenches around the Pacific Ocean.

Interpretation of Marine Sediments

The ocean basins are the major receptacles for the products of rock and mineral weathering on land. These products are transported as particles or in solution to the oceans, although they may pass through several environments on their way. After deposition and over time, these sediments become sedimentary rocks. By studying the characteristic compositions, structures, and textures of sediments being deposited today, we can understand the environments and processes that have led to the formation of sedimentary rocks. This is of practical significance because we know that economically valuable materials such as oil and gas, metal ores, and sand and gravel occur in certain environments. For example, we can look for sedimentary deposits that are good sources of oil and gas, such as organic-rich deltaic muds, and also look for deposits that will ultimately be good **reservoir rocks** for oil, such as the sand layers in shelves and in deltas. In contemporary environments we can examine the areal extent and pattern of the deposits using acoustic methods (see Exercise 6), the sediment characteristics observed from cores and grab samples, and the assemblages of biological remains that characterize the deposits. To identify these deposits in the geologic record we can use a variety of methods, including drilling wells and doing geophysical surveys, that give us samples of the rocks and yield a three-dimensional picture of the forms of the deposits and their internal structures.

Sediments deposited on the sea floor are not yet at the end of their journey. Deep currents can resuspend the fine muds much as windstorms on land move sediment as dust storms or sandstorms. Animals living in and on the bottom sediments constantly dig and burrow and stir the deposits in their search for food; therefore, the original layering of the settling particles is usually destroyed by this constant mixing. Only where the bottom water is devoid of oxygen because of rapid utilization of the oxygen do we see the original layering preserved. Examples of both of these conditions are shown in

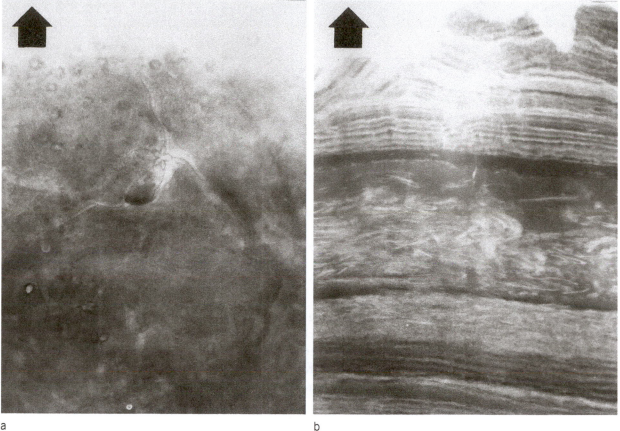

a   b

Figure 5-6 (a) Excellent slab photograph example of the activity of burrowing organisms as they mix bottom sediments. This sample was deposited in depths of about 100 meters off Southern California. There was enough oxygen in the bottom waters to support animal populations. The x-ray shows a starfish in its burrow and a small clam. The sediments have been thoroughly stirred and mixed many times. [Courtesy D. S. Gorsline, University of Southern California.] (b) Print of an x-ray photograph of a slab from a core of modern deep-water sediments in Santa Barbara Basin off southern California. Water depth is 600 meters. Because there is no oxygen in the bottom water, no animals live on or in these sediments, and so the annual additions of clays are preserved as thin layers. The thickness of layers is a function of the amount of rainfall each year and resulting changes in the river delivery of sediment to the sea floor. At a depth in the core of a few centimeters there is a disturbed zone that indicates that the surface sediment has moved relative to the lower sediment. This slippage also explains the irregular broken surface of the core.

Figure 5-6. In oxygen-free (**anoxic**) areas the organic matter deposited in the sediments also tends to be preserved, and these types of deposits may be eventual oil **source rocks.**

Material carried to the sea cannot accumulate indefinitely. If dissolved substances were not extracted from seawater, the chemistry of the oceans would change in a few million years, and if sediments were not converted into sedimentary rocks to become part of the land, the ocean basins would fill completely in a few hundred million years.

Thus a cycle recurs whereby rocks are weathered to form sediment that is transported to a depositional site to become rock again. Certainly this "recycling" of earth materials has occurred many times throughout earth history.

Sample Collection

Samples of the sea bottom can be collected by a number of different methods. Dredges of many designs, consisting of a simple frame with a net or

TABLE 5-2

Core descriptions

Core number	Description	Percentage of calcium carbonate in top 10 centimeters	Water depth (meters)
Ocean A			
1	Fragments of volcanic rock, some ooze	—	1500
2	10 meters of foraminiferal ooze, some volcanic fragments and volcanic ash in top meter	90	2000
3	10 meters of ooze, manganese nodules at top	80	3000
4	9 meters of ooze, manganese nodules at top	75	4000
5	3 meters of reddish clay and ooze over 3 meters of foraminiferal ooze	15	5000
6	8 meters of red clay	3	6000
7	9 meters of red clay	5	7000
8	1 meter of sand grading from fine at the top to coarse at the bottom, over 50 centimeters of muddy clay, over 60 centimeters of graded sand, over 3 meters of red clay	15	4500
9	6 meters of alternating gray clays and brown silts with some fine to coarse sand layers, one sand layer at base with gravel at bottom	—	1000
Ocean B			
10	4 meters of coarse sands and gravels	—	500
11	7 meters of silts with sand and gravel layers averaging about 40 centimeters thick	—	1500
12	9 meters of gray muds with fine sand and silt layers averaging about 20 centimeters thick	—	3000
13	10 meters of gray and brown clays and muds, several fine sand and silt layers 10 centimeters thick	6	4000
14	10 meters of brown mud with a few silt layers about 5 centimeters thick	62	3000
15	10 meters of brownish ooze	78	2000
16	Rock fragments and volcanic ash, rock is basalt	—	1500

bag attached, can be dragged along the bottom to collect materials. Weighted tubes can be dropped into the bottom and a core of the materials collected—a procedure that is much like pushing a cookie cutter into dough. Samplers with spring-loaded jaws can be used to grab a bite of the bottom materials.

The best samplers are those that collect an undisturbed core from unconsolidated bottom sediments, because cores display the sedimentary history of a given area. Layers of sediment repre-sent a finite period of time, and the kinds of sedi-ment tell us what the conditions were at the time of deposition. We can use the biogenous material to tell us about water temperature and, if radioac-tive materials are present, we can use various chemical methods to determine ages.

Sample Location and Analysis

A list of hypothetical core samples, and their de-scription, is shown in Table 5-2. Figure 5-7 is a

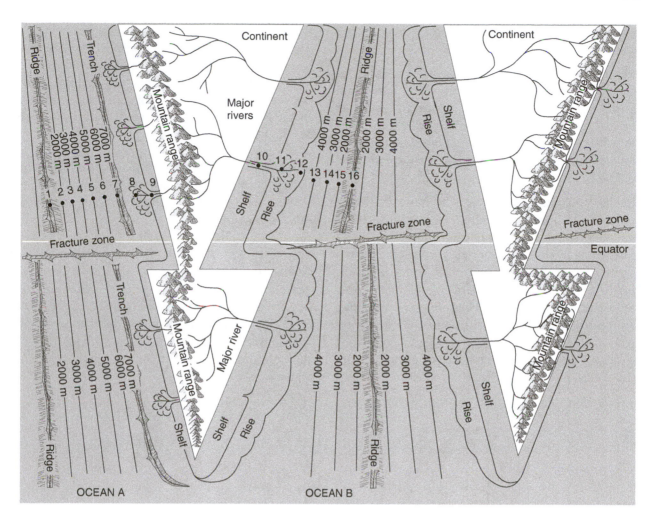

***Figure* 5-7** Map of hypothetical ocean basins showing locations of sea-floor cores. Contours indicate sea-floor depths in meters. Note that "Pacific-like" Ocean A has deep trenches at the base of the continental shelf, whereas "Atlantic-like" Ocean B has a continental shelf to continental slope to abyssal plain physiography.

hypothetical chart of oceans and continents in which we see the locations and approximate depths of these core samples, designated by sample number. Study both the table and the figure carefully, and refer to them as necessary in order to answer Questions 1–8.

DEFINITIONS

Anoxic. Without oxygen; denotes water or sediment from which free oxygen has been removed by oxidation of organic matter or by iron or sulfide in oxidation processes.

Antarctic bottom water (AABW). A water mass that forms around the margin of Antarctica where

formation of sea ice produces a cold, saline, dense water that sinks to the deep-ocean floor.

Carbonate compensation depth (CCD). Depth in the ocean where the solution of carbonate equals the supply of carbonate; it changes with time in response to changes in ocean biological production and ocean circulation driven by world climate changes.

Minerals. Naturally occurring, inorganic, crystalline substances with a definite range of chemical composition and reasonably definite physical properties.

Ooze. A deep-sea sediment consisting of at least 30 percent skeletal remains of microscopic floating organisms. The remainder is mostly fine-grained clay minerals.

Reservoir rocks. Coarse deposits of sand or gravel that have a high porosity after burial and lithification, and which, if oil and gas are generated nearby, can become saturated with hydrocarbons to form "pools," or reservoirs, much as sponges sop up water in their openings to hold a large amount of fluid.

Rocks. Aggregates of one or more minerals, rather large in area. The three classes of rocks are the following: (1) *Igneous rock.* Crystalline rocks formed from molten material. Examples are granite and basalt. (2) *Sedimentary rock.* A rock resulting from the consolidation or cementation of loose sediment that has accumulated in layers. Examples are sandstone, shale, and limestone. (3) *Metamorphic rock.* Rock that has formed from preexisting rock as a result of heat, pressure, or chemically active fluids. Examples are marble and slate.

Sediment. Loose fragments of rocks, minerals, or organic material that are transported from their source for varying distances and deposited by air, wind, ice, or water, or precipitated from the overlying water or form chemically in place. Sediment includes all the unconsolidated materials on the sea floor.

Slide (mass movement). A mass of sediment initially accumulated on a slope that eventually moves downslope as a mass; can be triggered by earthquakes or overloading.

Source rocks. Potential sources of oil and gas. Where conditions permit the accumulation of organic matter, the slow change in this material after burial can produce hydrocarbons. Typically, such high organic content is found in fine sediments that accumulated in areas of high biological production and low oxygen content in the receiving waters.

Exercise 5

Materials of the Sea Floor

NAME _____

DATE _____

INSTRUCTOR _____

Table 5-2 on page 50 gives sediment descriptions for the sea floor cores located on Figure 5-7 on page 51.

1. (a) Plot the data for percentage calcium carbonate (from Table 5-2) versus depth for cores 1–16 on the graphs below. Use an open circle for Ocean A data, and a closed circle for Ocean B data (it is not necessary to connect the dots for each ocean). If there is no $CaCO_3$ value, do not plot.

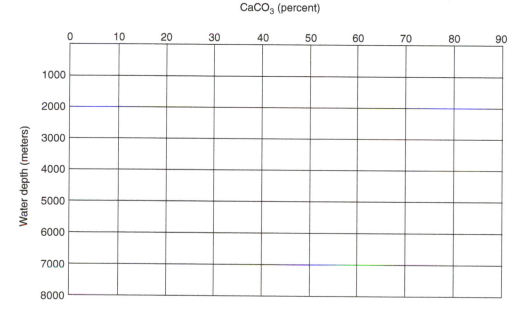

(b) Give an explanation for the sharp reduction in calcium carbonate below a depth of about 4000 meters in ocean A. _____

_____

_____

_____

_____

_____

(c) Can you provide an explanation for the difference in depth at which sharp reductions in calcium carbonate occur in Ocean A and in Ocean B? _____

_____

_____

_____

_____

2. On the graph below, plot a depth profile of each line of cores in Ocean A and Ocean B.
   On the profile, use colored pencils to indicate the *dominant* sediment type in each core. Use red for clays, blue for oozes, yellow for sand, silt, or terrigenous clays, and black for volcanic rock or basalt.

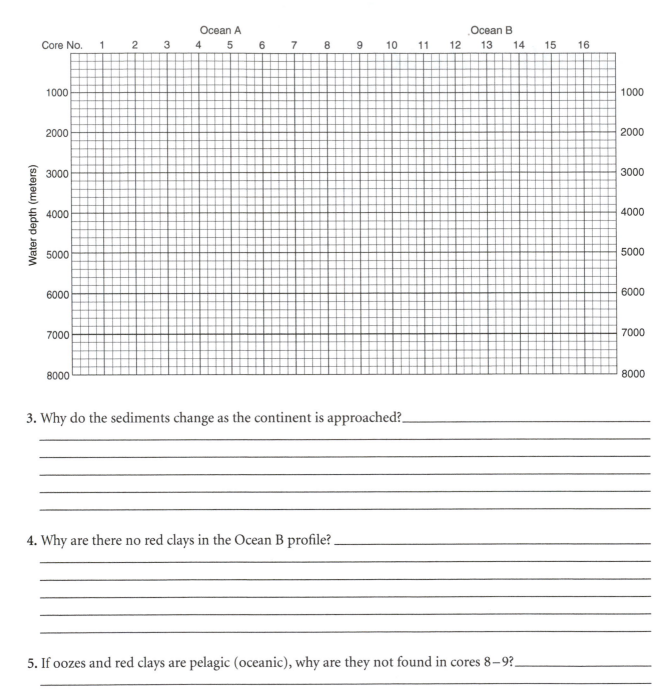

3. Why do the sediments change as the continent is approached?_____

_____

_____

_____

_____

_____

4. Why are there no red clays in the Ocean B profile? _____

_____

_____

_____

_____

_____

5. If oozes and red clays are pelagic (oceanic), why are they not found in cores 8−9?_____

_____

_____

_____

_____

_____

6. The sketch below shows core 5. How do you explain the presence of calcareous ooze below the red clay? (Hint: Remember the horizontal movement of the ocean floor caused by sea-floor spreading, and that sea floor sinks as it gets older.) _____

_____

_____

_____

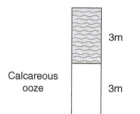

7. There are sands in core 14 in Ocean B at 3000 meters in depth, but none in core 3, which is also at 3000 meters, in Ocean A. Suggest the reason for this difference. _____

_____

_____

_____

_____

_____

8. Red clay on the deep-sea floor accumulates at an average rate of about 1 millimeter per 1000 years. Oozes accumulate at rates that are often ten times faster.
   (a) How much time would be required to deposit 1 inch of red clay? Show your work.

   (b) How much time would a core 10 meters in length represent if it contained 5 meters of ooze and 5 meters of red clay? Show your work.

9. Off the coast of California a special form of sea floor has developed because of the collision of a mid-ocean ridge and a continental plate. The resulting topography, shown in Figure 5-8, is called a *continental borderland* and is a checkerboard of deep basins, shallow banks, and islands. In the deep basins some representative sedimentation rates are as follows: San Pedro Basin, 54 centimeters per 1000 years; Santa Catalina Basin, 29 centimeters per 1000 years; San Nicolas Basin, 12 centimeters per 1000 years. Study the figure below and explain the differences in sedimentation rates between these basins. _____

_____
_____
_____
_____
_____
_____

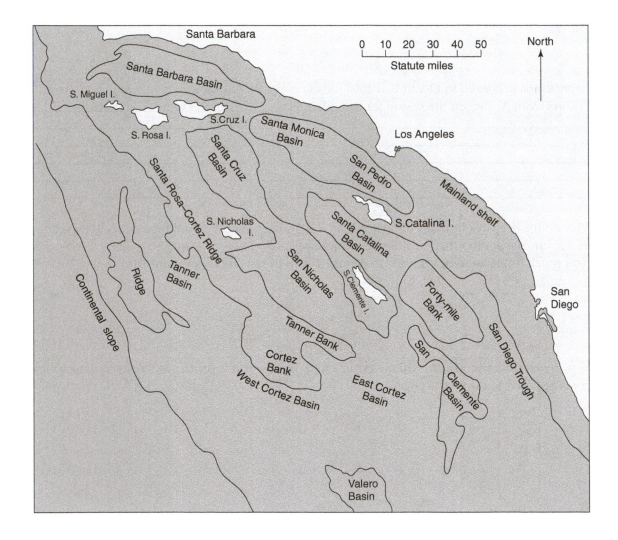

Figure **5-8** Simplified diagram of continental borderland off southern California showing major sedimentary basins and intervening ridges. Note that the channel islands represent portions of submarine banks that are above sea level.

Seismic–Reflection Profiling

OBJECTIVES:

■ To understand the principles of sound wave transmission and reflection in oceans and sea floors.

■ To understand the difference between low-frequency and high-frequency sound waves, and their use in resolving details of sea floor sediment layers.

■ To understand the principles of using side-scan sonar to visualize features of the sea floor.

In addition to observing and analyzing sedimentary materials, we need to know the geometry of those deposits and their architecture. We have seen in Exercise 1 how echo sounding can be used to collect information about ocean depth. We can also use this method to obtain sonic pictures of the forms of sediment deposits and of structures and features within these deposits. The discovery that using higher-power and lower-frequency sound sources could enable us to look at **reflectors** within bottom sediments occurred in the 1940s. This method was immediately adapted to both industrial surveying for economic deposits and research examining the processes by which sedimentary deposits are formed in a variety of ocean environments. The method gives us insight into the internal form of deposits much as x-rays and sound can be used by physicians to examine the inner structures of our bodies.

The basic principle of seismic-reflection profiling is the same as that of an echo sounder. Sound waves are emitted by a high-powered sound source, and the return echoes from reflectors within the sedimentary deposit are returned to the ship and recorded on the receiver (Figure 6-1). Sound wave **velocities** will depend on the nature of the material through which they are travelling: sound velocities in seawater range from 1450 to 1570 meters per second. When sound waves penetrate below the sea floor, they may penetrate through sub-surface sediment layers, or they may be reflected off a sediment layer and return to the sea surface. A sonic record from the reflections of these surveying arrays is shown in Figure 6-2. The circled numbers 1–4 designate the **reflections** and **multiples** in the photograph. The reflection marked 1 records a wave traveling from the source to the water surface to the receiver, and reflection 2 records a wave traveling from the source to the bottom to the receiver. Reflection 3 bounces off the bottom and returns to the surface, where it is reflected back down to the receiver. Finally, reflection 4 represents a multiple echo, since the pulse travels from the source to the bottom, then to the surface to be reflected back to the bottom, and then to the receiver. These different paths are shown schematically in Figure 6-3.

Sedimentary deposits can be many thousands of square kilometers in area and kilometers thick. We are also interested in features that have dimensions of meters or less. To see features at the full range of dimensional scales in ocean floor deposits, we need to use several different sonic methods. The **resolution**—the degree to which details are refined—of a sonic record depends on the **frequency** of the sound waves used. Frequency is defined as either the number of cycles per second (i.e., hertz) or thousands of cycles per second (i.e., kilohertz) emitted by the sound source. Sound

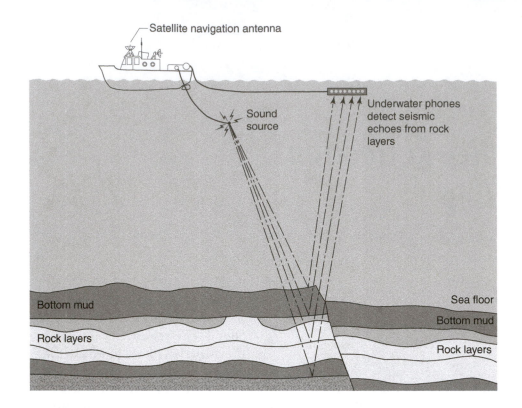

Figure 6-1 Seismic-reflection profiling. High-energy sound waves are bounced off subsurface sediments and rock layers to be picked up by underwater hydrophones or receivers. [From F. Press and R. Siever, _Earth,_ 4th ed. W. H. Freeman and Company. Copyright © 1986.]

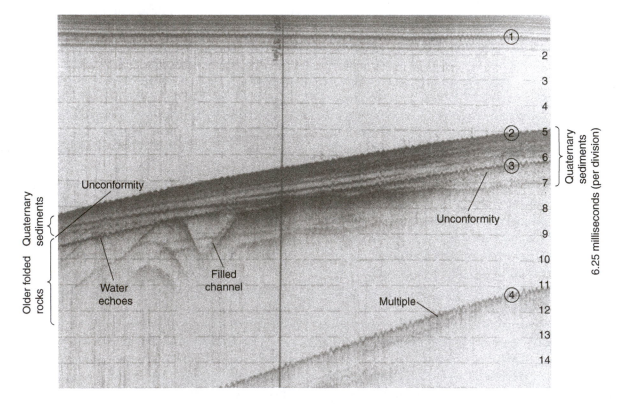

Figure 6-2 High-resolution seismic profile of shelf area off San Pedro Bay, California. The circled numerals 1–4 label echoes for the sound-wave paths shown schematically in Figure 6-3. [Courtesy Tom Nardin, University of Southern California.]

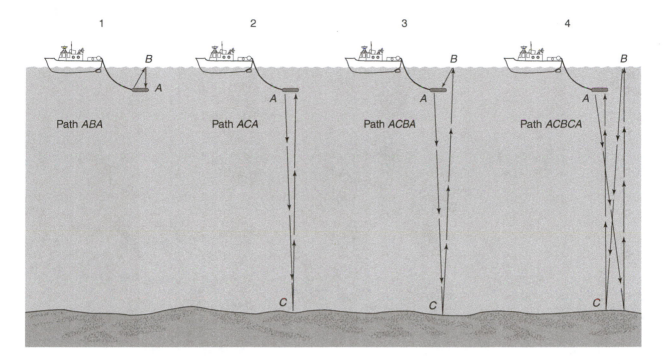

Figure **6-3** Schematic representation of a research vessel with a seismic-reflection profiling system in tow below the sea surface. The sound-wave paths shown are for the reflections recorded in Figure 6-2. The letter A refers to the sound source and receiver, B to the sea surface, and C to the surface of the ocean floor. Although the reflections shown are real, the angular relationships are distorted for purposes of clarity.

waves used in marine profiling range from relatively low-frequency (e.g., 2.5 kilohertz) to relatively high-frequency (e.g., 10 kilohertz). The frequency employed depends on the objective of the study. High-frequency waves yield high resolution records, but are rapidly dissipated as they move through water and sediment, and therefore limited in their penetration. In contrast, low-frequency waves penetrate more deeply, but cannot resolve features less than a few meters thick. Therefore, when using seismic-reflection profiling systems, several frequencies and arrays are often employed to obtain data at several levels of penetration and resolution.

We can also use sound to look at areas or swaths of the sea floor. Such devices use arrays of sound sources that send out sound waves laterally in a fan-shaped pattern so that echoes are received from various features in a broad band of the sea floor. This method is called *side-scan surveying* and a typical system is shown in Figure 6-4. These reflections give us a picture of the sea floor that is much like that from an aerial photograph of land

areas. An example of a side-scan survey is shown in Figure 6-5.

DEFINITIONS

Frequency. The number of complete waves that pass a given point in a second, or the number of complete vibrations per second. Short sound waves (high-frequency waves) yield rapid vibrations, whereas the vibration of long sound waves (low-frequency waves) is much slower. High-frequency waves give high-resolution records. Low-frequency waves give poorer resolution but much deeper penetration of the sea floor.

Multiple. A sound wave or pulse that follows a path from the source to the bottom, then to the sea surface, to the bottom again, and back to the receiver. The second round trip from the surface to the bottom and back is called the *first multiple* or *second echo*. As many as three or four multiples may be recorded if the energy source is large enough and the bottom acts as a strong reflector.

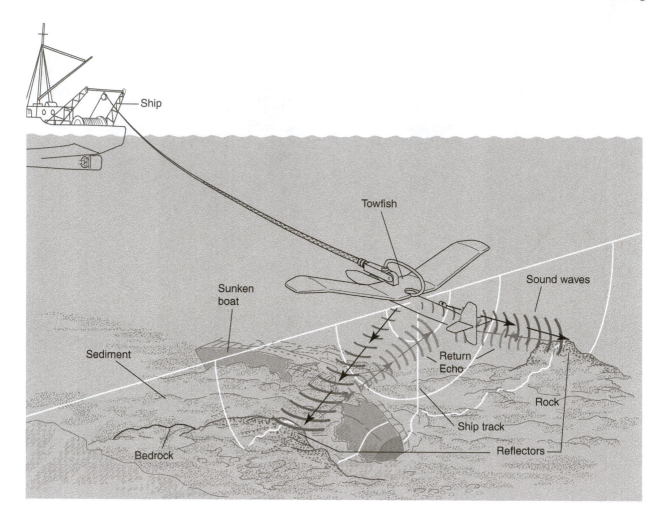

Figure **6-4** Typical side-scan sonar system. Note objects on sea floor that send return signals, or "echoes," back to the towfish.

Multiples may be recognized on the strip-chart record by the fact that each succeeding multiple or echo is displaced downward a distance equal to the water depth (see Figure 1-4).

Reflection. The return or bouncing off (the echo) of a wave from a surface back into its original medium. An example would be the reflection of sound waves that have traveled through seawater from a ship to the ocean bottom and back to the surface again. The same phenomenon occurs when sound waves travel through rock or sediment layers below the sea floor. Reflections are produced primarily by density differences between layers.

Reflector. A surface, usually a rock or sediment layer, that strongly reflects seismic (sound) waves.

Resolution. The degree to which details are defined on a record. High resolution means that fine detail is clearly shown. Resolution is dependent on the frequency of the sound.

10CCT02 Side Scan Sonar - Petit Bois Pass

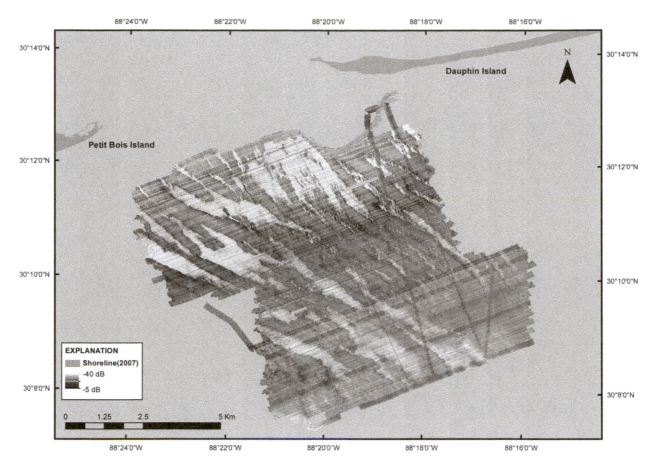

Figure 6-5 Sidescan sonar mosaic of the inner shelf of the Gulf of Mexico off Dauphin Island, Alabama, showing a migrating sandwave field on the seafloor. [U.S. Department of the Interior I U.S. Geological Survey.]

Seismic–Reflection Profiling

NAME _____

DATE _____

INSTRUCTOR _____

1. Figure 6-6 is a duplicate of Figure 6-2. The additional numbers 2 through 14 on the right side of this figure represents the *number* of one-way travel-times of 6.25 milliseconds (0.00625 seconds) required to travel from the source to the features present at each respective number. On the lines provided to the right of the figure, calculate these one-way travel-time data into meters below the source/receiver using the following relationship and data:

Depth (meters) = Total one-way travel-time × Velocity of sound in medium

Sound velocity in seawater = 1472 meters per second

Sound velocity in sediment = 1830 meters per second

For example, the total depth of water below the source/receiver at Number 5 will equal 46 meters (i.e., 5 one-way travel-times × 0.00625 seconds/one-way travel-time × 1472 meters/second). Note that to calculate the depths for each number greater than 5, you will need to caluclate the meters of sediment to that depth, and add this to the total water depth.

2. The depths calculated in Question 1 are depths below the source/receiver, not below the sea surface. Examine the nature of the reflections and multiples in Figure 6-3 and their one-way travel-time positions

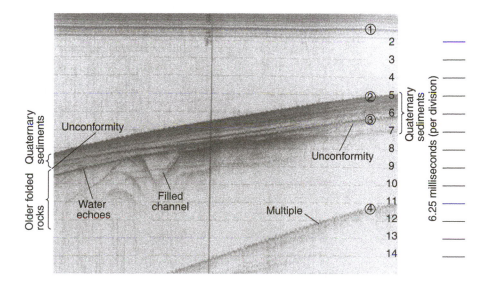

Figure **6-6** High-resolution seismic profile of shelf area.

63

in Figure 6-6. Using these relationships, calculate the approximate total water depth at number 5. Outline the method you used and show your work.

_____
_____
_____
_____
_____
_____
_____
_____
_____
_____

3. Based on your calculations in Question 2, at approximately what depth was the source/receiver towed?

4. The seismic profile in Figure 6-6 shows older folded sedimentary rocks covered by much younger, flat-lying Quaternary sediments. An erosion surface (unconformity) truncates the folded beds and represents the time during which the older rocks were eroded prior to deposition of the younger deposits.
  (a) What is the approximate thickness of the Quaternary sediments from the base of the filled channel to the sea floor?

_____
_____
_____
_____
_____

  (b) Assuming a rate of deposition of the young sediments of 40 centimeters per 10,000 years (a very fast rate), how many years did it take to lay down the young, flat-lying layers? _____ years.

_____
_____
_____
_____
_____

  (c) During what epoch of geologic time were these sediments deposited? (See Appendix B.)

_____
_____

5. In the side-scan sonar record of Figure 6-5, why are there smooth areas within and around the rocky outcrops? _____

_____
_____
_____
_____

6. What is the length of the side-scan record? _____

Optional Question

7. If the ship's course is 270 degrees true (due west), what is the approximate trend (strike) of the outcropping sedimentary layers? _____

*T*emperature and Salinity

OBJECTIVES:

■ To appreciate the importance of the vertical thermal gradient in the oceans and its importance in constraining the vertical motion of seawater.

■ To understand the physics of upwelling and its importance in coastal water temperature and climate moderation.

■ To discover what makes the oceans salty.

■ To learn how salinity varies in the oceans and the importance of rivers and ice in salinity variability.

Solar radiation is absorbed by seawater and stored as heat in the oceans. This absorbed energy may evaporate seawater and increase its temperature and salinity. Heating a substance usually causes it to expand and lowers its **density** (mass/unit volume) and when it cools, its density increases. The addition or subtraction of salts also causes seawater density to change: water of higher **salinity** will be denser than lower-salinity water. Pressure is another factor: as pressure increases with depth, so does the density of a water mass. Because high-density seawater tends to sink below average-density seawater, and low-density seawater tends to rise above average-density seawater, a change in density is one process whereby water motion can be generated. Therefore, we are interested in the distributions of both temperature and salinity, since these are two factors that determine vertical *thermohaline circulation* in the oceans.

Influences on Water Density

Of the three factors—temperature, salinity, and pressure—that affect water density, temperature changes have the greatest effect. We can see the importance of temperature as an influence on water density if we make a quantitative comparison of the effect on density of changes in salinity and temperature. We will exclude pressure effects from our investigation since they are very slight in water shallower than about 1000 meters. A change in salinity of one part salt per thousand parts of water (1 part per thousand, or 1 per mil, written as 1‰) has more effect on density than a change in temperature of 1°C. For instance, the density difference produced by a change in salinity of 1‰ is 0.001 gram per cubic centimeter; and the density difference produced by a temperature change of 1°C is, as a rule, between 0.00005 and 0.00035 gram per cubic centimeter. When we consider the surface waters of the oceans as a whole, however, we see that temperature is the more important factor, because its variations (ranging from −2° to 35ºC) are much greater than the salinity variations (which range only from 33‰ to 37‰).

The Thermocline

The permanent **thermocline,** a water layer within which temperature decreases rapidly with depth, acts as a density barrier to vertical circulation; that is, we may view this thermocline as the floor of the low-density, warm **surface,** or **mixed layer** and the ceiling for the cold, dense bottom waters (Figure 7-1). Over most of the earth, large-scale

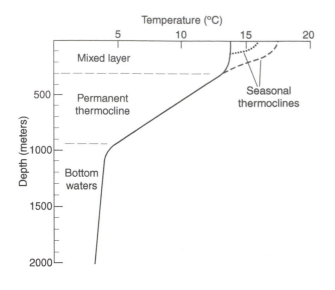

Figure 7-1 Thermoclines. During warm weather, if there are few storms, the upper 50–100 meters of the ocean warms, resulting in shallow seasonal thermoclines, as shown in the figure. The solid curve shows the winter condition in the mixed layer, and the permanent thermocline below. The dotted curve shows the seasonal thermocline that prevails after spring warming and that is present only in the upper mixed layer. The dashed curve shows the seasonal thermocline in extreme summer conditions, also present only in the mixed layer.

vertical movements of water between bottom and surface are inhibited by the strong contrast in density between these two layers. However, in the polar regions, surface waters are much colder, and therefore denser, so that little temperature variation exists between the surface waters and the deeper waters in these areas. Here vertical circulation takes place as surface waters sink to replenish deep waters in the major oceans.

Surface Temperatures

The distribution of surface temperatures in the major oceans is shown in Figure 7-2 (see also Color Plate 3). Points of equal temperature are connected by **isotherms.** Note that the isotherms tend to warp toward the equator on the east sides of the oceans, and poleward on the western sides. This is due to the major circulation pattern by which warm water is carried from the equator toward the poles on the west sides, and cooler water from the subarctic regions toward the equator on the east sides.

Water surface temperatures have important effects on coastal climates. Absorption of solar energy by seawater is one reason why coastal areas

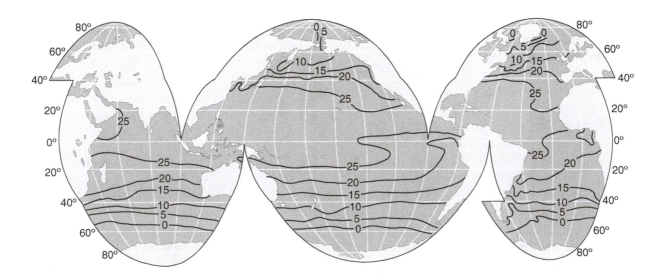

Figure 7-2 The distribution of surface ocean temperatures (in degrees Celsius) for the month of August. [Goode Base Map Series. Copyright © by the University of Chicago Department of Geology.]

have cooler temperatures in summer than inland areas. Re-radiation of stored heat from the ocean also causes coastal areas to remain warmer in winter than inland regions. Coastal currents also affect local climate. For example, San Diego, California, and Savannah, Georgia, have drastically different temperatures and humidities, even though they are both at 32° north latitude.

Upwelling

Another influence on surface temperatures in coastal waters of ocean basins is **upwelling,** the rising of cooler waters from greater ocean depths. In some coastal areas. the motion of the winds, combined with the Coriolis effect (see Exercise 9) pushes surface water offshore, and deeper water rises to the surface to replace it (Figure 7-3). Upwelling may be detected by surface temperature measurements, as it causes colder water to be found closer to shore, and isotherms warp upward toward the shore. Upwelling is particularly important along the coasts of Oregon, California, and Florida, where rising cool, nutrient-rich water masses support commercial and sport fishing in the nearshore waters.

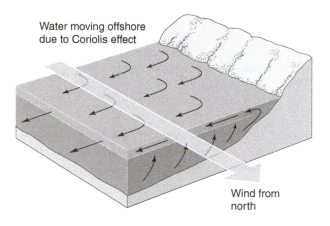

Figure 7-3 Coastal upwelling. In the Northern Hemisphere, Coriolis deflection causes surface waters to be pushed to the right of the wind direction. Off California, northerly winds (which blow *from* the North), combined with Coriolis deflection, force surface waters away from the coastline, causing deeper waters to upwell close to the shore.

Four other forms of upwelling are shown in Figure 7-4. Winds blowing consistently from the land outward over the ocean (offshore winds) may push surface waters offshore, and that water is replaced by upwelled water (Figure 7-4a). Upwelling may be caused by the movement of waters past a shoreline promontory ("obstruction upwelling"), causing deeper waters to rise in the bay past the point of obstruction (Figure 7-4b). One of the most biologically important regions in the oceans is the equatorial upwelling region (Figure 7-4c), in which separation of the surface waters by the trade winds brings cooler, nutrient-rich waters to the surface. Finally, the sinking of cold, dense surface waters (**downwelling**) close to the Antarctic continent causes upwelling of intermediate water farther away from the continent (Figure 7-4d).

Salinity

As noted, the density of seawater depends upon three properties, temperature, **salinity,** and pressure. As water cools, its density increases. Because water of high density tends to sink, and that of low density tends to rise above or settle below water that is at the average density of the oceans, the change in density is one process whereby water motion is generated. Therefore we are interested in the distribution of both salinity and temperature of water, since these are the two factors that determine the circulation that is caused by density changes.

The Distribution of Salinity

The oceans get their salt from the weathering and dissolution of minerals on land and from volcanic emissions. The mobile constituents of minerals are carried in solution by streams to the sea, where they accumulate and are recycled by various processes. Salinity is a "conservative" property; that is, it remains constant for the ocean as a whole for long periods of time, even though the local salinity varies within limits over the surface of the oceans. The average salinity for the oceans as a whole is 34.73 parts salt per 1000 parts water (34.73‰), but concentrations between 33.0‰ and 37.0‰ have been measured in the open ocean. High salinity or *dilution* is found only in coastal waters or in partially enclosed seas. Such extremes are due largely

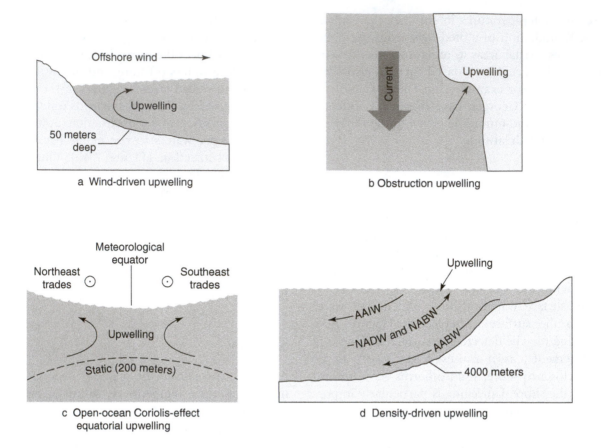

Figure 7-4 Diagrams of various kinds of upwelling. Figures 7-4a, c, and d are cross-sectional views, and Figure 7-4b is an aerial view. (a) Wind-driven upwelling. Offshore winds blow water away from the continent, and that water is replaced by upwelled water. (b) Open-ocean equatorial upwelling. Trade winds, combined with the Coriolis effect, transport water to the right north of the equator and to the left south of it, thus causing upwelling at the equator. The encircled dots indicate that the direction of the trade winds is toward the viewer. (c) Obstruction upwelling. A current moving past a headland or other obstruction will draw water away from the obstacle, and upwelling will occur. (d) Density-driven upwelling near the Antarctic continent. Cold, dense water sinks and replaces less dense water, and the less dense water upwells (see Exercise 8). AABW stands for Antarctic bottom water, AAIW for Antarctic intermediate water, NADW for North Atlantic deep water, and NABW for North Atlantic bottom water. The NADW and NABW are being upwelled.

to excessive runoff from the land, or to high evaporation rates and little mixing with other waters, as in the Red Sea and the Mediterranean Sea.

General variations in salinity are zoned from the equator to the poles. Values are low at the equator, highest in subtropical regions and at mid-latitudes, and lowest in the polar regions. The major processes responsible for this distribution are evaporation, precipitation, and mixing. Where evaporation exceeds precipitation, salinity values are high, and in areas of high rainfall, as at the equator, they are lower (Figure 7-5). The distribution of surface salinity in the major oceans for the

months of August is shown in Figure 7-6. Points of equal salinity are connected by **isohalines.**

Fresh water freezes at 0°C, but the addition of salt lowers its freezing point. For this reason, communities put rock salt on icy roads in the winter. At a salinity of 35‰ the freezing point of water is lowered to $-1.91°C$, almost 2° lower than fresh water. In polar regions, when freezing occurs, salt is expelled from the ice as it forms, and the freezing temperature of nearby water is depressed. This is because the water close to the ice becomes saltier and denser, and freezing stops until the temperature drops to the new freezing point.

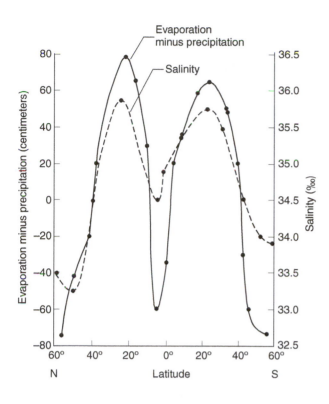

Figure 7-5 Distribution of surface salinity plotted against evaporation minus precipitation. [After S. Defiant, *Physical Oceanography*, vol. 1. Pergamon Press. Copyright © 1961.]

Determination of Salinity

The salinity of seawater is not a difficult property to determine. One reason is that regardless of the absolute concentration of salts in solution, the major dissolved constituents exist in virtually a constant ratio to one another. This fact was first recognized by Johann Forchhammer and later confirmed in 1884 by Wilhelm Dittmar, who carefully analyzed 77 samples collected on the *Challenger* expedition (1872–1876). Modern analytical techniques have enabled refinement of Dittmar's ratios; however, the importance of his work is not the accuracy of its numerical values, but rather its demonstration of the constancy of the ratios of about a dozen dissolved constituents (Table 7-1). In theory, if you determine the concentration of a major dissolved ion in a sample, you should be able to calculate the concentration of the other major constituents. In practice this is not quite so simple because of the analytical problems in distinguishing between several of the elements. Because chloride is the most common dissolved ion and one of the easiest to determine precisely, its concentration is determined, usually by a procedure known as the **Knudsen titration,** and from that measurement the salinity is calculated:

$$\text{Salinity (\textperthousand)} = 1.80655 \times \text{chlorinity (\textperthousand)}$$

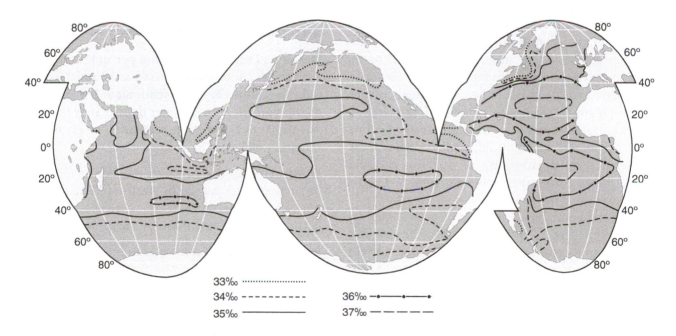

33‰	··········
34‰	– – – – –
35‰	————

36‰	–•——•——•
37‰	— — — —

Figure 7-6 Salinity distribution in the surface waters of the oceans in August. Notice the salinity differences between the Atlantic and Pacific coasts of the United States.

TABLE 7-1

Major dissolved constituents in seawater with a chlorinity of 19‰ and a salinity of 34.32‰

Dissolved substance	Concentration (grams per kilogram)	Ratio of dissolved salt to chlorinity (‰)	Percentage by weight
Chloride	18.980	0.99895	55.04
Sodium	10.556	0.55557	30.61
Sulfate	2.649	0.13942	7.68
Magnesium	1.272	0.06695	3.69
Calcium	0.400	0.02105	1.16
Potassium	0.380	0.0200	1.10
Bicarbonate	0.140	0.00737	0.41
Bromide	0.065	0.00342	0.19
Boric acid	0.026	0.00137	0.07
Strontium	0.013	0.00008	0.04
Fluoride	0.001	0.00005	0.00
Totals	34.482		99.99

Note in Table 7-1 that the salinity (34.32‰) calculated from the chlorinity of 19.00‰ is less than the salinity determined from the ratios of the elements to chloride ion (34.482). The reason is that bromine and iodine behave as if they were atoms of chlorine in the chemical analysis. However, the definition of salinity specifies that bromine and iodine should be converted to the chlorine equivalent (weight) and the carbonate converted to the oxide. When these mathematical manipulations have been completed, the chloride concentration increases, the carbonate and bromide contents are reduced, and the salinity calculated from a chlorinity of 19.00‰ and that determined from the ratios in Table 7-1 agree very well.

Another analytical method of determining the salinity of a salt solution is to measure the solution's ability to conduct an electrical current. **Conductivity** increases with increasing salt content, and this property of seawater may be measured with an electrical salinometer. At present, salinity determinations by high-precision conductivity measurements are more standard than chemical methods.

DEFINITIONS

Conductivity. The ability of a fluid to conduct electrical currents. The conductivity increases with increased salt content. Salinity may be determined by measuring the conductance of seawater with an electronic device called a *salinometer*.

Density. Defined as the mass per unit volume of a substance. In the metric system, the units for liquids are grams per cubic centimeter. For purposes of comparison only, seawater density may be taken as 1.0250 grams per cubic centimeter whereas fresh water is 1.000.

Downwelling. A downward movement (sinking) of surface water caused by onshore Ekman transport, converging currents, or when a water mass becomes more dense than the surrounding water.

Isohalines. Lines connecting points of equal salinity in the oceans.

Isotherm. A line connecting points of equal temperature, either at the sea surface or with depth.

Knudsen titration. The classical method for determining the chlorinity. The halides (chlorine, bromine, and iodine) are precipitated from a standard volume of seawater by a silver nitrate solution. The analysis is calibrated against a standard, or normal, seawater sometimes also called *Copenhagen water.*

Surface, or mixed, layer. The temperature zone of water above the thermocline where winds and currents mix the surface waters and convey heat downward.

Thermocline. Literally, a temperature gradient or rapid decrease of temperature with depth in a body of water; also the layer in which such a gradient occurs. The permanent thermocline in the oceans occurs between the levels of about 200 meters and 1000 meters, and separates an almost uniformly warm upper layer from very cold dense bottom waters.

Upwelling. The process by which water rises from a lower to a higher depth, usually as a result of offshore water flow. It is most prominent where persistent wind blows parallel to a coastline, so that the resultant Ekman transport moves surface water away from the coast.

Exercise 7

Temperature and Salinity

NAME

DATE

INSTRUCTOR

1. The table below lists temperature and salinity data for an oceanographic station off Point Conception, Southern California.

Depth (meters)	Temperature (°C)	Salinity (‰)	Depth (meters)	Temperature (°C)	Salinity (‰)
0	14.56	31.22	250	6.57	33.98
10	14.50	31.40	300	6.15	34.01
20	14.48	31.56	400	5.49	34.07
30	12.72	31.88	500	4.01	34.14
50	10.86	32.40	600	4.65	34.20
75	9.20	33.24	700	4.36	34.26
100	8.82	33.60	800	4.10	34.31
150	7.77	33.88	1000	3.51	34.41
200	7.10	33.94			

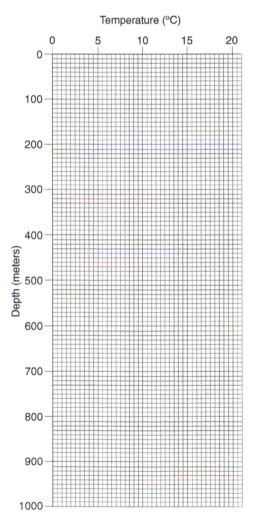

(a) Using the graph provided, plot the temperature data against depth.

(b) On your temperature-depth plot above, label the seasonal thermocline and the permanent thermocline.

(c) This profile was taken during the summer. What differences in temperature might you expect to see in the uppermost 200 m during the winter? _____

_____

2. The figure below contains oceanographic data for an east-west-trending line in the western North Atlantic Ocean off North Carolina.

Distance offshore (kilometers)	0	50	100	200	300	400	600
Temperature		15.50	18.20	23.92	28.42	24.95	24.60

Water depth (meters)

Shelf / Continental slope

Depth 0:
• 10.11 • 15.92 • 20.31 • 28.01 • 24.10 • 24.51
• 12.32 • 17.88 • 26.53 • 20.81 • 19.86

250: • 9.27 • 11.20 • 20.08 • 19.17 • 18.90

500: • 5.79 • 6.11 • 16.10 • 16.96 • 16.61

• 4.79 • 5.12 • 9.20 • 14.08 • 12.67

1000: • 3.81 • 3.92 • 5.82 • 8.61 • 7.62

1500: • 3.58 • 3.62 • 3.79 • 4.72 • 4.31

2000: • 3.16 • 3.18 • 3.22 • 3.46 • 3.40

3000: • 2.48 • 2.67 • 2.85 • 3.00

4000

(a) Contour the temperature data using a 2°C contour interval. Start with 16°C, then interpolate for warmer and cooler values.

(b) Thermoclines may be identified by subsurface isotherms that are more closely spaced than the overall spacing pattern. Locate and label the seasonal thermocline and the permanent thermocline in your contoured data.

(c) What process is indicated by the slope of the contoured isotherms above the part of the ocean floor labelled "Continental Slope?" _____

_____

_____

What is the result of this oceanographic process, in terms of movement of water masses?_____

_____

_____

_____

3. Figure 7-2 depicts the distribution of August surface temperatures in the major ocean basins. Refer to it for your answers to the following questions:

(a) Where is water of the greatest density formed? Explain your answer._____

_____

_____

_____

_____

(b) Is there any east-west difference in the surface temperatures of the equatorial Pacific Ocean? _____
What could cause these differences? _____
_____

(c) Explain the "bending" to the northeast of the isotherms in the North Atlantic Ocean? _____
_____

(d) How would a depth-temperature (BT) curve from the Arctic region differ from an equatorial depth-temperature curve? Sketch a curve for each region, and explain your answer. _____
_____
_____

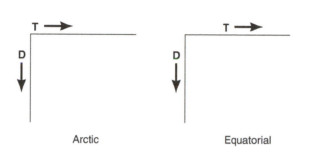

Arctic   Equatorial

4. Give the concentration of seawater with a salinity of 3.39 percent (parts per hundred) in the following units:
  (a) parts per thousand _____ (c) grams per kilogram _____
  (b) parts per million _____ (d) kilograms per metric ton _____

5. What is the salinity of seawater with a chlorinity of 19.65 per mil? _____

6. Study the isohalines on the map in Figure 7-6 and answer the following questions.
  (a) How does salinity vary in the Pacific Ocean from the South Pole to the equator to the North Pole?
_____
_____
_____

  (b) In the Indian Ocean, is salinity the same east and west of the Indian subcontinent? _____
_____

  (c) What reasons could explain the salinity pattern in the Bay of Bengal (eastern Indian Ocean)?_____
_____
_____
_____

  (d) Which is saltier, the Atlantic or the Pacific Ocean? _____
By what amount does their salinity differ? _____
How might you explain this difference? (Hint: Think of the major wind belts and dry areas of the globe, such as the Sahara Desert). _____
_____
_____
_____
_____

7. Figure 7-7 is a "gray-scale" reproduction of Color Plate 1. This figure shows salinity values for surface-water samples taken at the same time the satellite temperature image (Color Plate 1) was obtained.

(a) Contour the salinity values at intervals of 0.5 per mil (that is, at 34, 34.5, 35, and so on). Do the salinity contours parallel the Atlantic shoreline? _____
Can you explain the orientation of the isohalines relative to the shoreline? (Hint: Think of surface currents in this area.) _____
_____
_____

(b) Briefly explain the low-salinity stations near 38°N 72°W. _____
_____
_____
_____

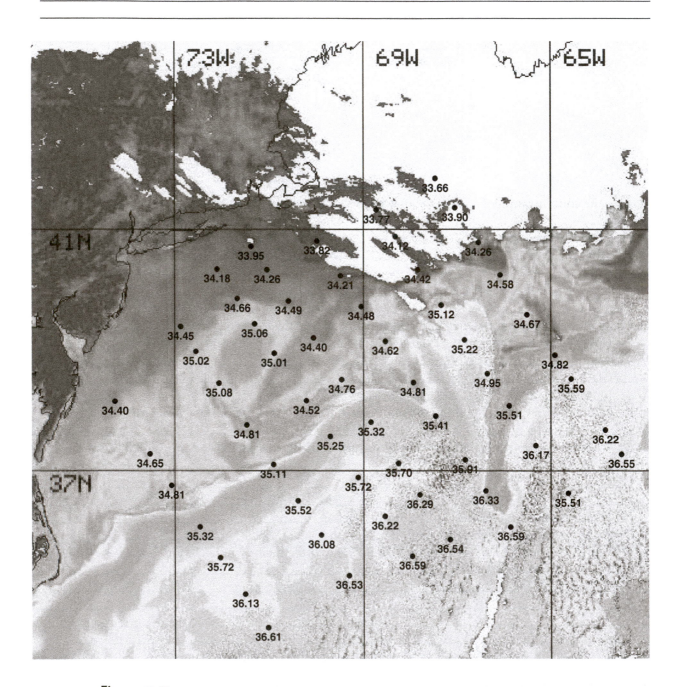

Figure 7-7 Surface salinity values (in parts per thousand) for 60 oceanographic stations off the Atlantic coast of North America, Fall 1982.

Water Masses and How We Study Them

OBJECTIVES:

■ To identify large masses of water that have a common origin or source area and to understand the oceanographic, biological, and meteorological significance of these masses.

■ To better understand the roles of temperature, salinity and density in the formation and transport of water masses.

■ To determine the temperature-salinity of a water sample and, by plotting these two parameter on a T-S diagram, identify its source area.

A **water mass** is a large volume of water that can be identified as having a common origin or source area. Water masses are formed by an interaction of water with the atmosphere or by the mixing of two or more bodies of water. Once formed, they sink to a depth determined by their *density* relative to the waters above and below them. The important determinants of density are *temperature* and *salinity*. All water in a water mass has about the same temperature and salinity.

Identification of Water Masses

Because water masses mix with the surrounding waters only very slowly, they tend to retain their original temperature and salinity. Thus the distinctive temperatures and salinities (and sometimes the oxygen content) of water masses make it possible to identify them. Their identification is important because it gives us information on their place of origin, on deep circulation, and on the rates at which waters of different densities mix.

Deep circulation—the motion of water at depth—is called *thermohaline* (temperature–salinity, hence density) *circulation* and is almost completely separate from that of the surface currents. Whereas surface circulation is largely in an east–west direction and moves warm water toward the poles, deep and bottom currents transport water in a north–south direction, returning cooler water along the meridians toward the equator. The cold water eventually returns to the surface to be reheated and returned to the poles by surface currents, or to mix with other waters and return to the depths. The velocity of thermohaline currents is very slow, about 1 centimeter per second, whereas surface currents are 10 or 20 times faster. Using the concept of **residence time**—the average time that a given substance (deep water in this case) remains in the ocean before being recycled—about 500–1000 years would be required to replace all the deep water in the Atlantic Ocean.

The identification of large water masses in the oceans is made possible by careful collection of oceanographic data. The most useful data for this purpose are temperature, salinity, and oxygen content. The extremes—that is, the maxima and minima for each of these parameters in a vertical column of seawater—are important for identification. Because the number of possible combinations of temperature and salinity is limited, it follows that only a reasonably small number of water masses are formed in the oceans. However, the density alone of a water mass is not sufficient for its identification because various combinations of temperature and salinity can produce the same density.

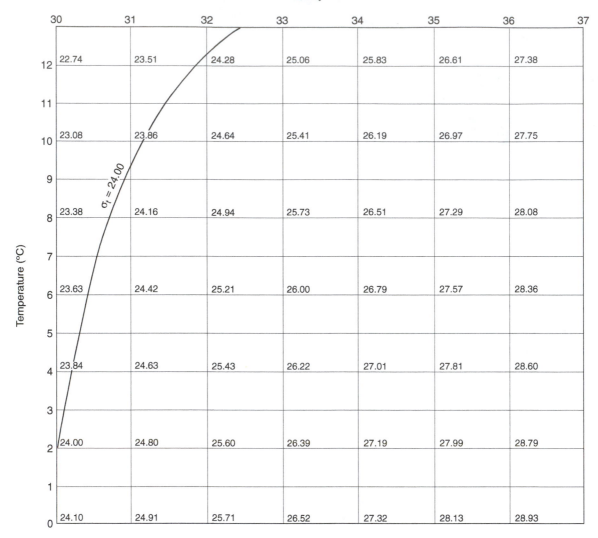

Figure 8-1 A temperature–salinity diagram showing σ_t values. Here the contour is $\sigma_t = 24.0$.

TABLE   8-1

Density factor, σ_t values for various temperatures and salinities

Temperature (°C)	Salinity (‰)						
	30	31	32	33	34	35	36
0	24.10	24.91	25.71	26.52	27.32	28.13	28.93
2	24.00	24.80	25.60	26.39	27.19	27.99	28.79
4	23.84	24.63	25.43	26.22	27.01	27.81	28.60
6	23.63	24.42	25.21	26.00	26.79	27.57	28.36
8	23.38	24.16	24.94	25.73	26.51	27.29	28.08
10	23.08	23.86	24.64	25.41	26.19	26.97	27.75
12	22.74	23.51	24.28	25.06	25.83	26.61	27.38

The Determinants of Density, and the Density Factor

Temperature, salinity, and pressure are the determinants of density, which is measured in grams per cubic centimeter. Inasmuch as the density of seawater is always greater than 1.0 gram per cubic centimeter (the density of fresh water), and never as great as 1.1 grams per cubic centimeter, it is more convenient to use a **density factor,** symbolized by the Greek sigma and subscript "t," σ_t. The density factor most commonly used takes into account temperature and salinity but ignores pressure, and is written as follows:

$$\sigma_t = (\text{density} - 1) \times 1000$$

Thus a seawater sample with a density of 1.02594 would have a $\sigma_t = 25.94$. Note that the mathematical manipulation involved in changing density to σ_t is simply dropping the 1 and moving the decimal point three places to the right.

Temperature–Salinity Diagrams

If we plot the density factors for a series of combinations of temperature and salinity on a **temperature–salinity** ($T-S$) **diagram,** we see that contours of equal density, **isopycnals,** are curved lines

(Figure 8-1). For example, water with a salinity of 32.00 parts per thousand (‰) at a temperature of 10°C has a σ_t of 24.64 (density = 1.02464). This value has been plotted in the $T-S$ diagram in Figure 8-1 where the 10°C and 32‰ lines intersect. Note the isopycnal in the figure. Every point on this line has a σ_t of 24.00; for example, this line crosses the 12°C temperature line at a salinity of about 31.85‰. Thus a water type with a temperature of 12°C and a salinity of 31.85‰ has a σ_t value of 24.00 (density = 1.0240 grams per cubic centimeter). Table 8-1 gives σ_t values for the range of temperature and salinity conditions commonly

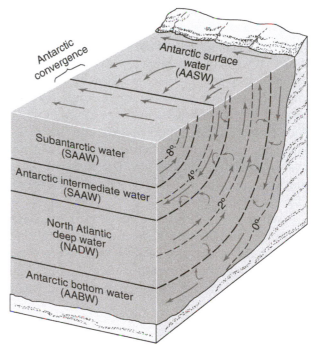

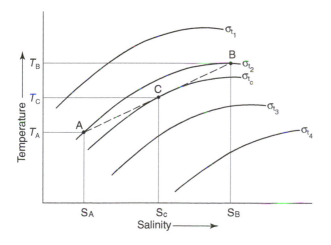

Figure 8-2 A temperature–salinity diagram showing the simple mixing of two water types, A and B, to form water mass C. Note that the density of C is greater than that of either of its end members A or B. The density increases from σ_{t_1} to σ_{t_4}.

Figure 8-3 Complex thermohaline and surface current circulation around Antarctica (north is to the left, and south is to the right in this figure). Sea ice formation next to the Antarctic continent causes surface water to cool and become more salty, forming Antarctic bottom water (AABW), which sinks and flows northward along the ocean floor. Eastward-blowing winds move Antarctic surface water (AASW) to the north (remember that in the Southern Hemisphere, Coriolis deflection is to the *left*), and cause upwelling of warmer salty water (NADW). Mixing of NADW with AASW forms Antarctic intermediate water (AAIW), which sinks below less dense surface waters at the Antarctic convergence. Generalized isotherms show temperatures characteristic of the water masses. [After V. G. Kort, "The Antarctic Ocean." Copyright © 1962 by Scientific American, Inc. All rights reserved.]

found on the open ocean. Note that the highest density, 1.02893 grams per cubic centimeter, is yielded by the water type with the highest salinity, 36‰, and lowest temperature, 0°C.

From Figure 8-2, we can see that the mixing of two **water types** of the same density, but of different temperatures and salinities, will produce a water mass that is denser than the two that originally mixed. This mixing process is known as **caballing.** In the figure, water types A and B, which have the same density, mix to form water mass C. When mixed in equal quantities, $T_C = (T_A + T_B)/2$, and $S_C = (S_A + S_B)/2$, but σ_{tC} is greater than $(\sigma_{tA} + \sigma_{tB})/2$, where T denotes the temperature, S the salinity, and σ_t the density factor. In general, the temperatures and salinities that result from mixtures of water types can be computed by simple averaging, whereas density cannot. Caballing produces intermediate water masses (500 − 1500 meters), deep-water masses (1500 − 4000 meters), and bottom-water masses in the oceans. Figure 8-3 shows these water masses in the South Atlantic off Antarctica. Surface-water masses and types are generally formed by direct interaction and exchange between sea and air. The different water masses found in the North Atlantic Ocean are listed in Table 8-2.

When you plot the $T-S$ values for a real water station (as in Question 4), you will note that they are in a stable position relative to each other because their densities increase as we go deeper in the column. The relative degree of stability is shown on a $T-S$ diagram by the angle the $T-S$ curves make with the contours of density. If the curves cut across the lines of equal density at large angles, the water column changes density rapidly. This means that minor changes in density will not cause water to sink or rise a great distance, and that the water column is therefore very stable. If the curves are at very small angles to the density contours, then the water changes density only slightly with increasing depth, so that minor changes in temperature and salinity (density) will cause water to move to different depths rapidly.

Also, a $T-S$ diagram in a well-known area of the oceans can be used to correct or find data points that are in error. The incorrect data appear as isolated points on the diagrams outside of the areas in which the water conditions are usually found.

DEFINITIONS

Caballing. A mixing of two water masses of identical *in situ* densities but different temperatures and salinities. The resulting water mass then becomes denser than either of its components.

Density factor (σ_t). A convenient numerical value for manipulating density data: $\sigma_t = (\text{density} - 1) \times 1000$. If density is 1.02386 then σ_t is 23.86.

Isopycnal, or isopycnic line. A line of equal or constant density.

TABLE   8-2

Water masses in the North Atlantic Ocean

Water mass	Source	Identifying characteristic
Antarctic bottom water (AABW)	Weddell Sea	Bottom temperature minimum
North Atlantic deep water (NADW)	Area near Greenland	Intermediate salinity maximum, may show intermediate temperature maximum, low oxygen content in South Atlantic
Surface-water masses	Regional	Variable, generally warm
Mediterranean intermediate water (MIW)	Mediterranean Sea off Turkey	High salinity and temperature tongue at intermediate depths

Residence time. The mean length of time a quantity of a given substance spends in the ocean before being removed or recycled.

Temperature–salinity (*T–S*) diagram. A diagram on which the characteristics of large masses of water are identified by plotting salinity and temperature data. Salinity is plotted with increasing values toward the right, and temperature is plotted with decreasing values downward. Depth at which each sample has been taken is usually indicated along the curve.

Water mass. A large volume of seawater that can be recognized as having a common origin. Water masses may be formed by interaction between air and sea or by mixing of two or more water types. A water mass is characterized on a *T–S* diagram by a group of values that may be plotted as a curve or a straight line.

Water type. A homogeneous mass of water having well-defined temperature and salinity characteristics. It appears as a single point on a *T–S* diagram.

Exercise 8

Water Masses and How We Study Them

NAME

DATE

INSTRUCTOR

1. Below is a duplicate of Figure 8-1 in which the σ_t value 24 has been plotted. Plot the values from $\sigma_t = 25$ to $\sigma_t = 28$ on the $T-S$ diagram. You may wish to approximate σ_t values between those listed for specific temperatures on salinity isohalines.

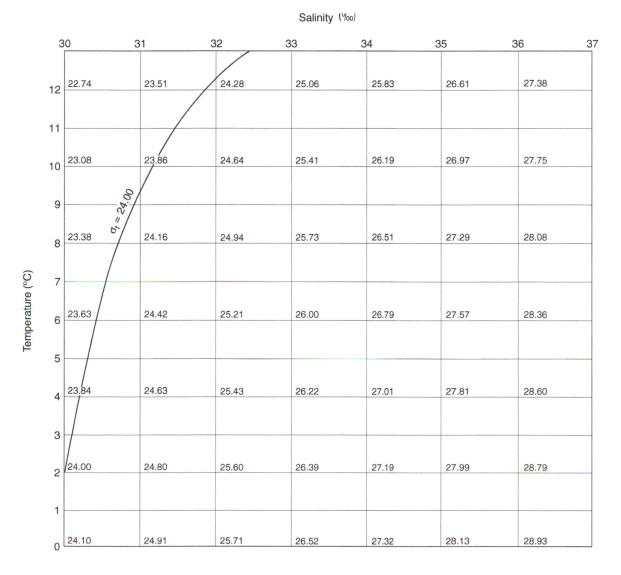

Salinity (‰)

2. On the 25.00 contour, plot a water type A where the isopycnal (density contour) crosses the 33‰ salinity line, and a water type B where the isopycnal crosses the 3°C temperature line. Plot a point in the middle of this mixing line. Note that it falls below and to the right of the 25.00 σ_t contour and is therefore denser. This midpoint is the temperature and salinity of a 50–50 mixture of water types A and B. The line itself represents all possible mixtures of water types A and B and thus the water mass that would form from such a theoretical mixture.

   (a) What is the range of T and S that defines the water mass produced by mixing water types A and B in all proportions? _____

   (b) What is the density of the 50–50 mixture of water types A and B? _____

   (c) What is the temperature of the 50–50 mixture of water types A and B?_____

   What is the salinity of the 50–50 mixture of water types A and B?   _____

3. The table below lists temperature and salinity data from an oceanographic station in the eastern North Pacific Ocean. Plot these data on the T–S diagram you contoured in Question 1, and connect the data points with straight lines.

Depth (meters)	Temperature (°C)	Salinity (‰)
0	9.1	32.4
10	9.1	32.4
20	9.1	32.5
50	9.1	32.5
100	8.2	33.6
200	7.1	33.9
300	6.2	34.0
400	5.5	34.1

   (a) How deep is the mixed layer, and in what portion of the curve is the water most stable? Use the table showing depths and also use the plot you made. _____

_____

   (b) Crosshatch the portion of the curve where the water mass is most stable.

4. The following are oceanographic data are from a typical station in the North Atlantic Ocean at about 20°N latitude.

Depth (meters)	Temperature (°C)	Salinity (‰)
100	16.0	36.1
200	13.0	35.7
400	11.0	35.4
500	9.0	35.3
600	8.0	35.2
850	11.8	36.9
950	11.2	36.6
1200	9.9	36.3
2000	4.0	35.0
2200	3.5	34.9
2500	2.0	34.8
3000	0.0	34.7
4000	− 1.2	34.6
5000	− 1.9	34.5

Salinity (‰)

Plot the data on the blank $T-S$ diagram above and draw a line connecting all points in order of depth. Using the diagnostic parameters provided in Table 8-2, name the major water masses represented in the diagram. Circle the two high-salinity points at depth.

5. The velocity of the southward flow of North Atlantic deep water through the Atlantic Ocean has been estimated at 6×10^6 cubic meters per second.
   (a) If the volume of the Atlantic Ocean is 3.24×10^{17} cubic meters, how long would it take to circulate all the Atlantic water through the deep water? Show your work. _____

   (b) If the oceans are at least 3 billion years (3×10^9) old, and we assume a constant rate of mixing, how many times has the ocean been "stirred" through the deep water? Show your work. _____

Surface Currents

OBJECTIVES:

■ To understand and recognize the major movements of surface water in the oceans.

■ To see how currents influence climate by redistributing energy from the sun and stored heat on earth.

■ To understand the roles that wind, gravity, and the earth's rotation play in determining the direction and velocity of ocean currents.

■ To understand the physics of upwelling and its importance in coastal water temperatures and climate moderation.

Surface oceanic circulation is the result of several processes, including wind stress acting on the water surface and differences in density due to solar heating. If we assume that observed **current** systems are simply the result of wind stress, then they should very closely follow the major wind belts on the earth (Figure 9-1). In fact they do, but you will note that there is a slight clockwise divergence (to the right) from the wind direction in the Northern Hemisphere and a counterclockwise one in the Southern Hemisphere. This is due to the earth's rotation, and its influence on ocean currents was not appreciated until 1835 when Gaspar de Coriolis, while studying equations of motion in a rotating frame of reference (the earth), discovered what is now called the **Coriolis effect** or, not quite accurately, the Coriolis force. The word "force" is misleading because the effect is due not to force or acceleration, but to the earth's eastward rotation, and any moving object not attached to the earth's surface shows the apparent deflection described above. The amount of deflection depends on the velocity of the object and its latitude. It is zero at the equator and maximum at the poles, and fast-moving objects are deflected more than slow-moving ones. It should be noted that the Coriolis effect has no influence on the energy of motion and modifies only direction, that is, to the right in the Northern Hemisphere and to the left in the Southern Hemisphere.

Major Currents of the World's Oceans

Figure 9-2 is a map of the major wind-driven currents in the world's oceans. Note that in the Northern Hemisphere the currents form large clockwise-rotating gyres or rings. Note also that the gyres are not exactly centered in the ocean basin but are slightly offset to the west. Thus the currents along the western sides of the northern oceans are narrow, fast, and deep, whereas those on the eastern sides are wide, shallow, and sluggish. As an example, the Kuroshio Current in the western Pacific Basin has about six times the flow of the California Current on the eastern side, but is only one-fourth as wide. Velocities within the Kuroshio Current can attain 10 kilometers per hour, whereas the California Current generally moves with a velocity of less than 2 kilometers per hour. This phenomenon is known as *westward intensification,* and is due to the earth's rotation and the necessity of balancing or conserving angular momentum. The same phenomenon may be observed in the Canary–Gulf Stream current system in the Atlantic Ocean.

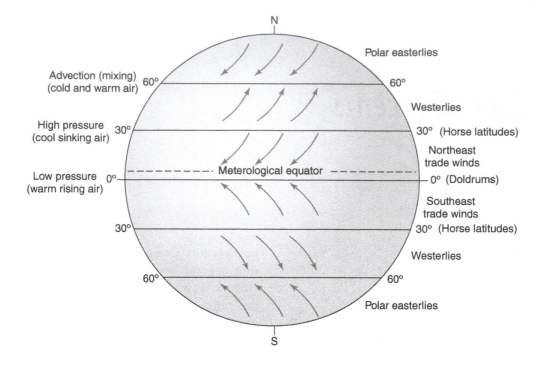

Figure 9-1 Major wind belts on earth and zones of high and low pressure. Note that the meteorological equator is 5–10° north of the geographic equator.

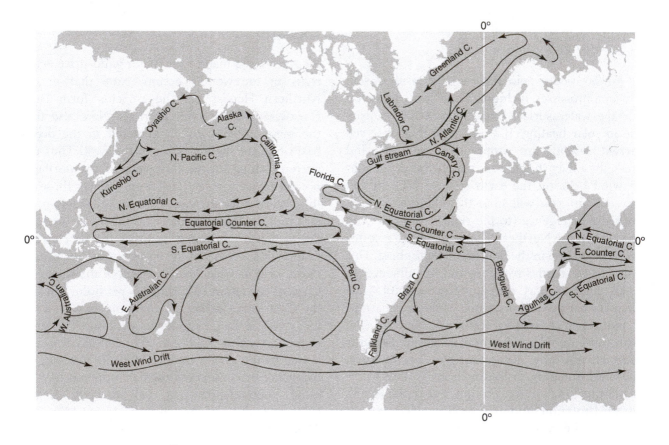

Figure 9-2 Major surface currents of the oceans of the world.

The currents in the southern oceans are essentially mirror images of those in the northern parts. They rotate in a counterclockwise direction, as predicted by Coriolis, and show westward intensification. Note that the plane dividing the northern and southern circulation lies a few degrees north of the equator. This plane, known as the *meteorological equator*, is slightly north of the *geographic equator* due to the physical differences between the Northern and Southern hemispheres (see Figure 9-1). In the Northern Hemisphere continental landmasses occupy a greater area than the oceans, and winds are slowed by turbulence and friction as they flow over irregular terrain. In contrast, the Southern Hemisphere is ocean dominated, with less land area and huge expanses of open ocean (see Figure 9-2). For these reasons, southern ocean waters are subjected to overall higher wind speeds. Sailors in the South Pacific Ocean were well aware of this, giving high southern latitudes such descriptive names as *roaring forties* and *screaming sixties*. The meteorological equator also marks the convergence between the northeast and southeast trade winds, and is a zone of rising warm air (low pressure), billowy water-laden clouds, and erratic winds. It is known as the *doldrums* and was the scourge of sailing vessels attempting to cross the equator.

The west-flowing North and South Equatorial currents cause water to build up in the western equatorial Pacific Ocean, and meteorological satellites have measured this ocean-surface topography. Periodic decreases in southeast trade-wind stresses on the ocean surface cause the bulge of warm surface waters to move eastward along the equator, resulting in a phenomenon called *El Niño*. We will discuss the effects of El Niño in Exercise 17.

Volumes of Water Flow

Major currents transport tremendous volumes of water. A flow unit called the sverdrup (Sv), named after Harald U. Sverdrup, a famous oceanographer, is used to indicate a flow of 1,000,000 cubic meters per second. The volumes of transport of water of the major currents of the Pacific Ocean are shown in Figure 9-3.

Data for the Atlantic Ocean are more difficult to obtain, but it is estimated that the flow of the Gulf Stream is at least 55 Sv, and for the Canary Current a range of 2–16 Sv is estimated. The westward intensification is clearly demonstrated in these figures.

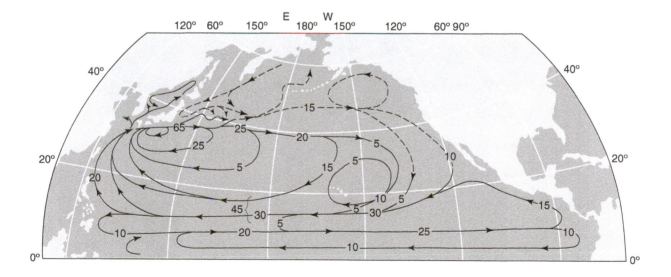

Figure **9-3** Transport chart of the North Pacific. The lines with arrows indicate the approximate direction of water transport above 1500 meters, and the numbers show transported volumes in sverdrups. Broken lines show cold-water currents; solid lines warm ones. [After Harald U. Sverdrup, Martin W. Johnson, and Richard H. Fleming, *The Oceans: Their Physics, Chemistry, and General Biology.* Copyright © 1970. Redrawn by permission of Prentice-Hall, Inc., Englewood Cliffs, New Jersey.]

Speed of Flow and Dynamic Topography

The greatest sustained speeds are found in the Kuroshio Current, which is known to move as fast as 3 meters per second. Speeds as great as 50 centimeters per second are reported in the westward flow near the equator and the West Wind Drift.

An ideal way to measure the velocity and direction of currents would be to establish current-meter stations in all parts and at all depths of the oceans. The exorbitant expense and difficulty in-

volved in mooring and maintaining such meters has limited the use of such direct methods to relatively small areas. The common procedure used to determine large-scale current motion is to measure the distribution of density (therefore pressure) in the oceans or of a particular current and convert the readings into profiles or sections showing horizontal density changes or anomalies. The patterns of these anomalies reflect the deflection of the water surface from horizontal and enable the investigator to construct a map showing **dynamic topography,** or irregularities in the sea surface. The

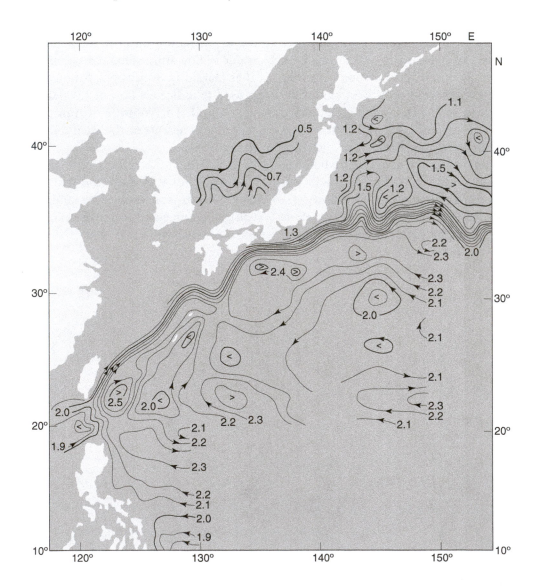

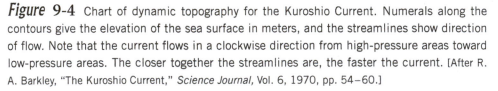

Figure 9-4 Chart of dynamic topography for the Kuroshio Current. Numerals along the contours give the elevation of the sea surface in meters, and the streamlines show direction of flow. Note that the current flows in a clockwise direction from high-pressure areas toward low-pressure areas. The closer together the streamlines are, the faster the current. [After R. A. Barkley, "The Kuroshio Current," *Science Journal*, Vol. 6, 1970, pp. 54–60.]

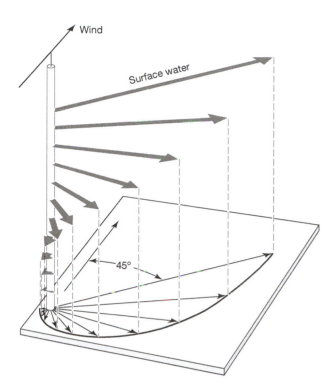

Wind

Surface water

45°

Figure 9-5 The Ekman spiral in the Northern Hemisphere. Surface water is deflected at a 45° angle to the right of the wind direction, and successively deeper layers of water are deflected progressively farther to the right of the layer immediately above. This results in a weak current of deeper water in a direction opposite the direction of surface winds.

topography so illustrated in dynamic meters reflects horizontal pressure gradients, and it is then possible to obtain useful approximations of real currents (to within about 15 percent). To do this, a series of temperature and salinity measurements are made to a depth of about 1 kilometer at a number of locations (called *stations*). From these data the horizontal pressure gradients can be computed and a chart of dynamic topography prepared. If we contour values of equal dynamic topography and assume that only the Coriolis effect applies, then the flow will theoretically parallel the contours. When we put arrowheads on the contours pointing in the proper direction, they indicate the direction of flow and are called *streamlines*. Figure 9-4 is such a chart, showing dynamic topography and streamlines for the Kuroshio Current in the western Pacific Ocean. The spacing of the streamlines gives us an idea of the speed of flow. Where they are closely spaced, the horizontal pressure gradient is large and the flow will be fastest.

The Origin of Dynamic Topography

How are the irregularities in the sea surface formed and maintained? To answer this question we must examine the interaction between wind, gravity, and the Coriolis effect. In 1905 V. W. Ekman provided a theoretical explanation that laid the cornerstone for all subsequent studies of wind-driven currents. Observers had noted that icebergs in the northern oceans moved 20–40° from the wind direction. Ekman showed that, given a steady wind and homogeneous sea, surface water will flow 45° to the right of the wind direction in the Northern Hemisphere and 45° to the left of it in the Southern Hemisphere. However, this surface layer, a few meters or tens of meters thick, sets in motion an underlying layer that is deflected farther to the right (in the Northern Hemisphere), but with lesser velocity because of friction. Subsequent layers downward are set in motion, each being deflected farther to the right until at depth there is a weak current flowing in the opposite direction. This rotation of the wind-driven current with depth is known as the **Ekman spiral** (Figure 9-5). The base of the wind-driven water column varies with wind velocity and persistence, but is usually taken to be at a depth of 100 meters. This theory itself was revolutionary, but more important is the fact that the average direction of flow throughout the entire water column is 90° to the right of the wind direction in the Northern Hemisphere, and 90° to the left in the Southern Hemisphere. This net motion of water at right angles to the wind is known as *Ekman transport*.

Now we are able to see how dynamic topography develops. The northeast trades and the westerly winds in the Northern Hemisphere push water into a bulge near the center of the current gyres, owing to Ekman transport (Figure 9-6). This bulge of warm surface water is unstable, and the water attempts to return to a level stable condition with a uniform distribution of warm water near the surface and cooler water below. As a water parcel starts to flow downslope by gravity (to a region of lower pressure), the Coriolis effect deflects its motion to the right after it reaches an appreciable velocity. As it continues downslope it is turned farther to the right because the Coriolis effect is continually operating to the right of its motion. Once it has turned 90° it cannot turn farther

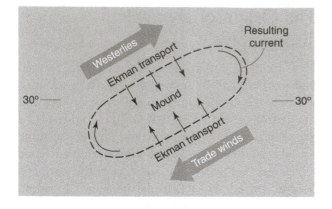

Figure 9-6 The creation of an oceanic high-pressure area at 30°N by prevailing winds. [After J. Williams, J. J. Higginson, and J. D. Rohrbough, *Sea and Air: The Naval Environment.* Copyright © 1968, U.S. Naval Institute Press.]

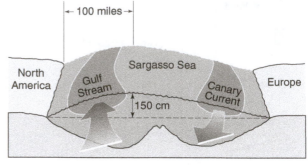

Figure 9-8 The mound in the North Atlantic Ocean between the Gulf Stream and the Canary Current. The Sargasso Sea is a region of warm, clear surface water, deep blue in color and containing large quantities of sargassum, or seaweed. [After J. Williams, J. J. Higginson, and J. D. Rohrbough, *Sea and Air: The Naval Environment.* Copyright © 1968, U. S. Naval Institute Press.]

without flowing uphill. At this point there is a perfect balance between the Coriolis "force" and gravity, and the water parcel continues to move around the mound parallel to the contours of dynamic topography, or streamlines (Figure 9-7). If it turns slightly downhill it gains speed, and the Coriolis effect deflects it to the right again. In this way a balance is reached between Coriolis "forces" and pressure forces (gravity). Currents generated in this manner are called **geostrophic** (earth-turned) **currents.** A schematic of this phenomenon for the North Atlantic is shown in Figure 9-8. It can be ap-

preciated that once the mound of water is elevated by the wind it is not relieved by geostrophic flow, because the water simply moves around the hill. However, in nature this is not exactly true, because some friction occurs in fast-moving currents and a water parcel eventually spirals down back to a low. It is said that if the wind were to stop, then the ocean would become perfectly flat in about 3 years and surface currents as we know them would cease to exist; however, density differences would continue to cause movement at depth.

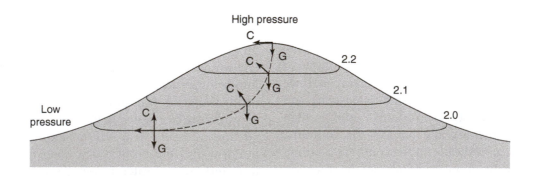

Figure 9-7 Diagram of the forces generating geostrophic currents in the Northern Hemisphere. Note the perfect balance that is achieved between the Coriolis effect (C) and gravity (G) as a water parcel moves around the mound parallel to the contours. Numerals represent the height in dynamic meters. [After W. Anikouchine and R. Sternberg, *The World Ocean.* Copyright © 1973. Redrawn by permission of Prentice-Hall, Inc., Englewood Cliffs, New Jersey.]

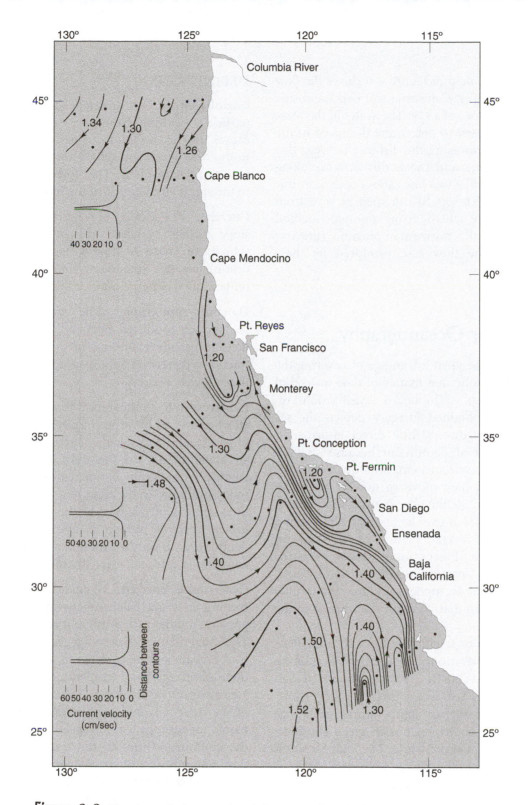

Figure 9-9 The dynamic topography of the sea surface during the summer of 1939 off California and Baja California, relative to an arbitrary level surface at depth. Note that the direction of flow alternates toward and away from the coast. Contours represent the height of the sea surface in dynamic meters, with contour intervals of 0.02 meters. Surface current speed can be closely approximated by measuring the distance between adjacent contours, then comparing this measure to one of the scales on the left side of the figure. [After Harald U. Sverdrup, Martin W. Johnson, and Richard H. Fleming, *The Oceans: Their Physics, Chemistry, and General Biology.* Copyright © 1970. Redrawn by permission of Prentice-Hall, Inc., Englewood Cliffs, New Jersey.]

The current map in Figure 9-9 shows the flow of surface waters off California and Baja California during the summer of 1939. The scales in the illustration can be used to determine the speed of the current: Simply measure the distance between two adjacent contours and move this distance along the gap between the two lines above each scale until they match. The speed can then be read from the scale. You are actually using a graphic method for calculating the horizontal pressure (gravity) changes and the flow rate produced by these changes.

Satellites for Oceanography

Satellites have the great advantage of covering the surface of the globe in a matter of days instead of months or years, as shipboard investigations require. Satellite-obtained imagery depicts the sea surface only; it conveys little direct information about the water at depth. But because satellites move so fast, they record data in a quasi-synoptic manner, that is, they move so quickly across an ocean basin that satellite observations may be considered simultaneous over large ocean regions, and give us a better overall picture of the ocean at a particular time. Furthermore, responses at the sea surface to perturbations in temperature, salinity, or dynamic topography are significantly briefer than those that occur in the interior of the ocean.

An example of satellite imagery is Color Plate 1. It is an Advanced Very High Resolution Radiometer (AVHRR) image obtained in November, 1982. This image is color-coded, showing warmer sea-surface temperatures in orange and yellow and cooler regions in green and blue. Cloud-covered areas are white. The Gulf Stream is clearly shown by the sharp boundary between warmer water to the southeast and cooler water to the northwest.

DEFINITIONS

Coriolis effect. An apparent "force" on moving particles resulting from the earth's rotation. It causes moving bodies to be deflected to the right in the Northern Hemisphere and to the left in the Southern Hemisphere. The "force" is proportional to the speed and latitude of the moving object.

Current. The motion of water as it flows down a slope, pushed by wind stress or tidal forces. The velocity or speed of flow is usually expressed in centimeters per second, or for fast-moving currents in meters per second or kilometers per hour.

Dynamic topography. The irregularities in the sea surface produced by wind or differences in density, usually expressed in dynamic meters or fractions thereof. This topography changes with time and the seasons.

Ekman spiral. A theoretical representation of the effect of a wind blowing steadily over a large body of water, which causes the surface layer to drift at an angle of 45° to the right in the Northern Hemisphere. Water at successive depths drifts more and more to the right in a spiral fashion until at some depth, known as the *base* of the wind-driven current, motion is essentially zero. This depth depends on the duration and velocity of the wind but is approximately 100 meters.

Geostrophic current. A current defined by assuming an exact balance between the horizontal pressure gradient (density) and the Coriolis effect. The usual manner of deriving geostrophic currents is to prepare a chart of dynamic topography based upon observations of temperature and salinity for a number of oceanographic stations. The direction of the current is indicated by the contours of dynamic topography, and its speed by the spacing of the contours. Although the underlying assumptions are only approximately correct, the direction and speed computed by this method are very close to the direction and speed actually observed.

Exercise 9

Surface Currents

NAME

DATE

INSTRUCTOR

Report

1. Based on the information in Figure 9-2, describe the following currents as either relatively warm or cold, and fast or slow (based on western intensification).

Current	Relative temperature	Relative speed
California		
Kuroshio		
West Wind Drift		
Benguela		
Gulf Stream		
Canary		
Labrador		
Agulhas		

2. Why is the Equatorial Countercurrent in the Atlantic Ocean so poorly defined in comparison with the same current in the Pacific Ocean? _____
_____
_____
_____

3. (a) What is the only current that completely mixes water between the Atlantic, Pacific, and Indian Oceans? _____
(b) Why was it so difficult for mariners to leave the South Atlantic Ocean and enter the South Pacific Ocean, using sail-powered ships? _____
_____
_____
_____

4. The "America's Cup" yacht races were traditionally held off Newport, Rhode Island (41°N 71°W) on the U.S. East Coast. When the "America's Cup" was won by Australia, the races were moved to Perth, on the southwestern coast of Australia (32°S 115°E).

95

(a) What change might you expect to see in water temperature off Perth, Australia, compared to the former race location off Rhode Island? _____

_____

What changes in wave height (which is largely controlled by surface wind strength) would you expect off Australia? _____

_____

In current strength and direction? _____

(b) The America's Cup races have also been held in the eastern Pacific, off San Diego, California (32°N 117°W). Would you expect ocean conditions off San Diego to be closer to Newport, Rhode Island, or closer to Perth, Australia? Explain your reasoning. _____

_____

_____

_____

_____

5. Using the scales on the California Current map (Figure 9-9), determine the maximum and minimum current velocities off Point Conception. Use the scale closest to the speed to be determined. _____

_____

6. A temperature cross-section from San Francisco, California, to Honolulu, Hawaii, is shown in Figure 9-10. The section is through the California Current and the eastern North Pacific central water to a depth of 500 meters. Surface water temperature, salinity, and temperature-depth profiles were taken at each of 21 stations. The table below shows surface temperature and salinity readings. The California Current is

Station number	Surface salinity (‰)	Surface temperature (°C)
1	33.20	11.2
2	32.90	12.5
3	33.10	14.8
4	33.20	14.5
5	—	15.0
6	33.50	15.8
7	33.45	15.2
8	33.40	15.7
9	33.41	16.0
10	34.65	17.0
11	34.95	17.5
12	35.10	18.5
13	35.20	18.9
14	35.20	19.0
15	35.20	19.8
16	35.25	20.5
17	35.10	22.0
18	35.15	22.4
19	35.10	22.8
20	35.10	23.2
21	35.15	23.7

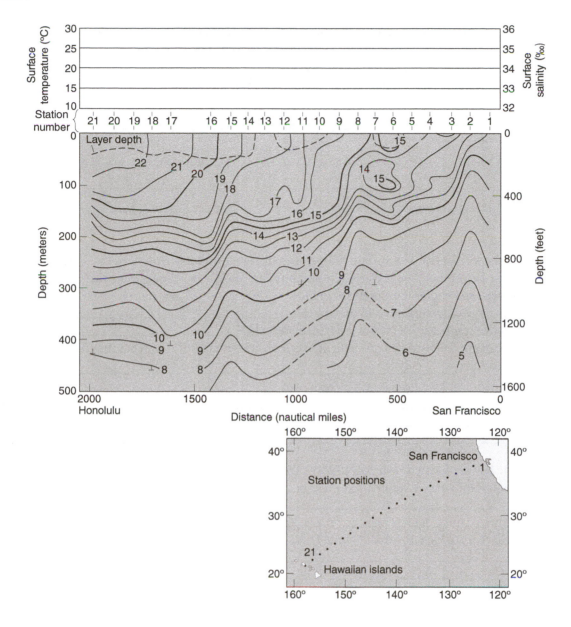

***Figure* 9-10** Cross section of temperature to a depth of 500 meters from San Francisco to the Hawaiian Islands. Temperatures in degrees Celsius are indicated along the isotherms. The symbol ⊥ shows the depth of the bathythermograph drop. (From *Fishing Information*, U.S. Department of Commerce, National Oceanographic and Atmospheric Administration, April 1972.)

recognizable by colder surface waters, low salinities, and irregular isotherms. Eastern North Pacific central water is much warmer, with salinities in excess of 34.8‰. A transition, with salinities in the range of 34.0–34.8‰ exists between the two water masses.

(a) Plot the surface salinity data at each station on the graph above the section, using the salinity scale at the right-hand side. Use color and connect the points for each station with a line.

(b) Plot the surface temperature at each station in the same manner, using the temperature scale at the left-hand side. Use a different color and connect these data points with a dashed line.

(c) How far from San Francisco does the region of eastern North Pacific central water appear in the section? _____

A surface mixed layer can be recognized by vertical isotherms. To what depth does the surface mixed layer extend? _____

(d) How wide is the California Current in this section? _____

(e) Fill in the table below, to compare these three surface water masses:

Current	Isotherm shape	Surface water temperature	Salinity
California Current	_____	_____	_____
Transition Zone	_____	_____	_____
Eastern North Pacific central water	_____	_____	_____

7. Color Plate 1 shows AVHRR sea-surface temperature imagery for a portion of the western North Atlantic Ocean for November, 1982. Warm temperatures are shown in orange, intermediate temperatures in yellow and green, and cooler temperatures in blue.

(a) What is the large current shown in the image? _____

What is its relative temperature (warm/intermediate/cold)? _____

(b) Does this current follow a straight path across the entire image? If not, what might cause lateral "meanders" to occur? _____

_____

_____

_____

_____

(c) What is the feature located near 39°N 72°W? What is the relative temperature of this feature, compared to adjacent surface waters? _____

_____

How might this feature have formed? (Hint: Think about surface current meanders). _____

_____

_____

_____

_____

8. Previously, this same region of the North Atlantic was shown in Figure 7-7, which also depicted oceanographic stations (and salinity data) sampling surface waters.

(a) Compare your salinity isohalines (from Exercise 7-7) to the surface temperature data shown in Color Plate 1. Which provides more detailed information, the AVHRR image or the surface salinity stations of Figure 7-7?_____

(b) Are there lateral meanders evident in Color Plate 1 that are not readily apparent in the salinity data of Exercise 7-7? If the AVHRR data are more detailed, why would it be desireable to send oceanographic vessels to this region of the ocean? _____

_____

_____

_____

_____

_____

Tides

OBJECTIVES:

■ To understand the roles that gravity and centrifugal force play in influencing the heights of oceanic tides throughout a lunar month.

■ To understand why tides arrive at different times and why they display different magnitudes and frequencies in coastal and deep water.

■ To understand how the shape of the coastline influences the tides.

■ To know what tidal data planes are and why are they important to navigation.

To the casual observer the most obvious change in the level of the sea is that of the tides. They are caused mainly by the moon and the sun exerting forces on different parts of the rotating earth. The tidal wave, or bulge, is the result of *gravitational attraction* and *centrifugal force*, which act in combination to produce a regular variation in water level in the course of a day.

Tide-Producing Forces

Consider only the earth–moon system. Although the moon appears to revolve about the earth, the two bodies are actually rotating about a common center of mass. They are held together by gravity and kept apart by an equal and opposite centrifugal force. Thus, on the side of the earth closest to the moon, the tide-producing force is gravitational, whereas on the opposite side centrifugal force dominates.

We can illustrate this point by showing the water motion resulting from only the horizontal tide-generating forces (Figure 10-1a). When the moon is over the equator, water is drawn by gravitational attraction toward the side nearest the moon, and toward the opposite side by centrifugal force, so that high tides with a low tide belt in between result. Because the forces are of equal magnitude symmetrically about the equator, two high and two low tides of equal magnitude should be experienced, at least in theory, at any given latitude on the earth. As the moon shifts north or south of the equator (that is, when it is at north or south declination), the forces are as shown in Figure 10-1b. A point at the equator is still subject to highs and lows of equal magnitude, but points at higher latitudes will experience strong diurnal inequalities; in other words, they will experience high tides of unequal heights, or perhaps only one high tide.

Tide Levels and Datum Planes

Because the ocean basins vary in size and shape, and because land masses interfere with the tidal bulge, the tides do not assume a simple regular pattern. Although a purely mathematical solution of the tidal phenomenon is still beyond the limits of marine science, it is possible to predict the tide level at least 1 year in advance by careful analysis of tide records from stations at which observations have been made for long periods of time. Most of the averages are based on at least 19 or 20 years of records, and quite accurate prediction is routine.

The tide level is usually measured in reference to a local base level, or **datum** plane, which is an

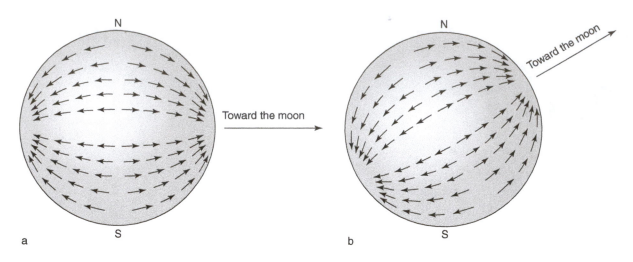

Figure **10-1** Tide-producing forces. The arrows represent the magnitude and direction of the horizontal tide-generating forces on the earth's surface. The force pulling away from the moon is the centrifugal force produced by the rotation of earth and moon about their common center of mass. (a) When the moon is in the plane of the earth's equator, the forces are equal in magnitude at the two points on the same parallel of latitude on opposite sides of the earth. (b) When the moon is at north or south declination, the forces are unequal at such points and tend to cause an inequality in the two high waters and the two low waters of a tidal day. [After N. Bowditch, *American Practical Navigator.* Hydrographic Office Publication No. 9, U.S. Naval Oceanographic Office, 1966.]

average of many years' observations. The typical datum in the United States is mean lower low water (MLLW), which is the average of the lowest tide each day. Another common datum is mean low water (MLW). This is the average of all low-tide levels at the station, but is not as safe a point of reference for navigational purposes as mean lower low water, since at least half of the lows during a month will be lower than the datum. Other datum planes are mean sea level (MSL), mean high water (MHW), and mean higher high water (MHHW). Inasmuch as mariners depend on the charted depths, and since these are established in reference to the tide datum, it is obvious that the best datum will be the lowest normal level that the tide will reach. The relationship between levels of the sea and datum planes for the Orange County coast of California are shown in Table 10-1.

Types of Tides

Three major types of tides can be recognized on the basis of frequency of occurrence and symmetry of the tidal curve. **Diurnal tides** occur once

daily, meaning that there is one high and one low tide of about equal amplitude, or height, in the course of a **tidal day.** A tidal day is 24 hours and 50 minutes long because the moon, which exerts the greatest tidal influence, advances 50 minutes each day in its orbit around the earth. **Semidiurnal tides** occur twice daily and are also of about equal height. **Mixed tides,** also known as *irregular semidiurnal tides,* occur twice daily but exhibit two highs and two lows of significantly unequal height. The type of tide that occurs on a given coast and its variation in height depend on a number of factors. Among them are the shape of the basin in which the tide occurs, natural oscillations of the water (seiches) within the basin, declination of the sun and moon, and relative position of the sun and moon. Tides on the East Coast of the United States are representative of the semidiurnal type, whereas those on the West Coast are mixed tides. However, either coast may exhibit both types at certain times of the year. Diurnal tides typically occur in partially enclosed basins such as the northern Gulf of Mexico, the Java Sea, and the Gulf of Tonkin off the Vietnam–China coast (Figure 10-2).

TABLE 10-1

Sea levels and datum planes for the coast of Orange County, California

Sea level	Datum mean sea level (feet)	(meters)	Datum mean lower low water (feet)	(meters)
Highest tide	4.8	1.5	7.5	2.3
Mean higher high water	2.6	0.8	5.3	1.6
Mean high water	1.9	0.6	4.6	1.4
Mean sea level	0.0	0.0	2.7	0.8
Mean low water	− 1.8	0.5	0.9	0.3
Mean lower low water	− 2.7	0.8	0.0	0.0
Lowest tide	− 5.2	1.6	− 2.5	0.7

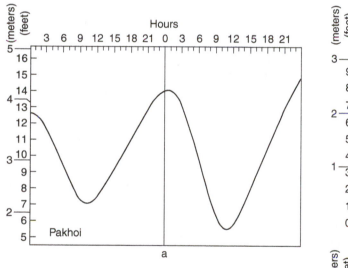

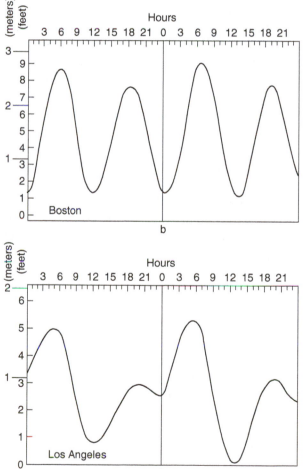

Figure 10-2 Types of tides from the Atlantic and Pacific ocean basins: (a) diurnal type; (b) semidiurnal type; (c) mixed type.

Monthly Tidal Cycles

Tides are also identified by their **tidal range:** those having, in the course of a lunar month, the largest difference in level between high and low are called **spring tides;** those having the smallest range of the month are the **neap tides.** Spring tides occur twice monthly, at or near the time of new moon and of full moon. At these times, tides are at their highest and lowest levels in relation to their mean level. Perhaps the best datum to use for navigational purposes would be the average of all low spring tides or mean lower springs. Neap tides—the tides of lowest range (the lowest high tides and the highest low tides)—are also influenced by the lunar cycle and occur twice a month at or near the first quarter and third quarter phases of the moon (the half moon).

Other fluctuations in the tidal range occur in response to the elliptical nature of the moon's orbit. When the moon is at a point closest to the earth (the *perigee*) once every 27.5 days in its orbit about the earth, the range of the tide is increased; when it is at the point farthest away from the earth (the *apogee*), smaller tidal ranges occur, other factors being equal.

The sun, too, causes tides, the solar influence being slightly less than half that of the moon. The solar tidal cycle occurs in the course of a 24-hour period and is not synchronous with the lunar tidal period of 24 hours and 50 minutes. However, when the sun, moon, and earth are in alignment, as they are at new and full moons (Figure 10-3a,b), the lunar and solar components reinforce one another and spring tides result. When neap tides occur, at first- and third-quarter moons, the sun–earth–moon system forms a right angle and the tide-producing forces are greatly diminished (Figure 10-3c).

The typical tidal curves given for various localities in Figure 10-4 show the three major types of tides and the effects of proximity and alignment of the moon and sun. Note the tidal curve for New York for September 22–26. The tidal range is high because the sun, moon, and earth are lined up; the sun and moon are at the equator, producing a high degree of symmetry, and the moon is at perigee, causing the higher spring tides to occur at this time rather than at the new moon phase when the moon was at apogee. The

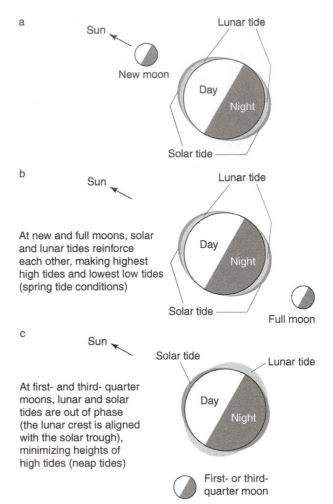

Figure 10-3 The relative positions of earth, moon, and sun determine the tidal ranges during the lunar month. The highest tidal ranges (spring tides) occur at new and full moons, when the lunar tidal crest is superimposed on the solar tidal crest. Minimal tidal ranges (neap tides) occur at first- and third-quarter moons, when the lunar tides "cancel out" the solar tides—the lunar tidal crest is superimposed on the solar tidal trough, and the lunar trough and the solar crest coincide. [After F. Press and R. Siever, *Earth,* 4th ed. W. H. Freeman and Company. Copyright © 1986.]

other curves may be explained in the same way, except the one for Port Adelaide, where the solar and lunar tides are about equal, so that they nullify one another at neap tides (at the other localities the lunar tide-producing force is about twice that of the sun).

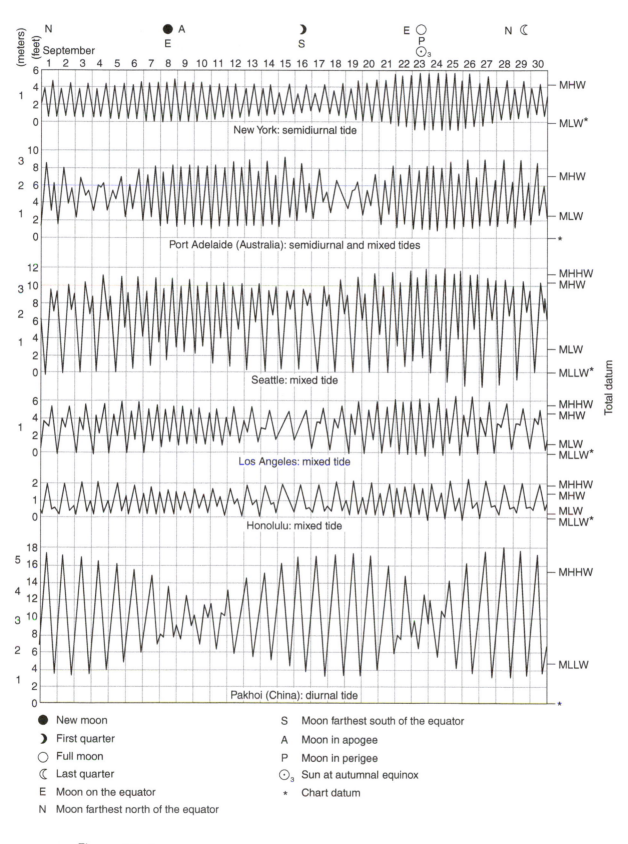

Figure 10-4 Examples of tide records for several tidal types. [After N. Bowditch, *American Practical Navigator.* Hydrographic Office Publication No. 9, U.S. Naval Oceanographic Office, 1966.]

Unusual Tides

There are places in the world where tidal ranges exceed 10 meters and may reach as much as 16 meters. These occur in bays or harbors open to the ocean that are very long compared to their depth. Natural oscillations, known as *seiches,* in these basins cause water to slosh back and forth much like the waves you can create when walking with a coffee cup. Physicists refer to these waves as *stand-*

ing waves, or *forced oscillations,* because the water stands first high and then low through one cycle. If the fundamental period (one up-down cycle) of the tidal basin or harbor is equal to the Earth's tidal period (12 hours and 25 minutes), then resonance results, and extreme tidal ranges may occur, as positive reinforcement "stacks" the crests of waves resonating in the basin. A simple "desktop" demonstration of resonance can be obtained with an empty soda bottle. Blowing air gently across the open top of the bottle may produce a loud tone due to resonance of the air vibrating inside the bottle.

Figure 10-5 shows the way in which the natural frequency of oscillation of a tidal basin and the tidal frequency may be in phase to produce resonance. This constructive reinforcement creates large tidal ranges—as great as 16 meters in the Bay of Fundy, 10 meters in the upper reaches of the Gulf of California, and 10 meters at Anchorage, Alaska. In some narrow funnel-shaped estuaries a *tidal bore* develops. This phenomenon, an abrupt solitary wave that moves upstream with the incoming tide, can be quite dangerous, since bores range in height from a few inches to as much as 25 feet. The most famous is on the river Severn in England; about 4 feet high, it can pass an observer on the run (Figure 10-6). Bores occur on the Amazon River (up to 25 feet high), on the Knik and Turnagain arms of the Cook Inlet, and on the Petitcodiac River at the head of the Bay of Fundy, also noted for its extreme tidal range during spring tides.

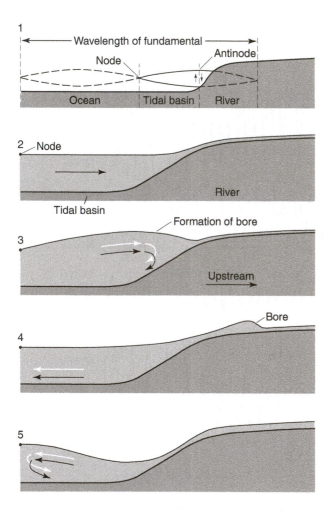

Figure 10-5 Oscillation, or seiche, of the water in a tidal basin. The time it takes the standing wave to make one complete oscillation is the fundamental period. The node is a point of little or no vertical movement of water; the antinode is a point of maximum vertical movement of water. If the tide rises fast enough, a bore is formed in the river mouth. The effect of constructive reinforcement, or resonance, is shown in panels 2–5. [After D. K. Lynch, "Tidal Bores." Copyright © 1982 by Scientific American, Inc. All rights reserved.]

Storms and Water Level

In most coastal areas the wind may induce surface-water flow in the direction of wind motion and thus cause the water level to rise or fall above or below that level owing to astronomical tides. The term **wind setup** is used when this effect takes place in a lake or reservoir, and **storm surge** is applied to the same effect along the open coast. It is extremely important for the planning of engineering projects to know how much storm surge can be expected in a coastal area.

For example, storm surge in southern California is predicted to be about 1 m above highest tide levels; therefore engineering works should be constructed at least 3 m above mean lower low water (see Table 10-1). The amount of surge de-

Figure 10-6 Surfer riding the Severn River tidal bore wave, South East Wales.
[© The Photolibrary Wales/Alamy]

pends on the wind velocity, the length of open sea surface across which the wind can generate waves, and the depth of the water; surge is greater for shallow water. The influence of shallow water is the reason that storm-surge values are higher on the Gulf Coast than on the Atlantic Coast (and on the Atlantic Coast higher than on the Pacific Coast). Indeed, in 1900 a hurricane surge on the Gulf Coast Galveston, Texas, was so high that water levels rose 5 m above mean lower low water, inundating much of the coastal land; and in any year surges of 2–3 m feet above tide levels are not unusual along the southeastern coast of the United States.

Tsunamis and Tides

Tsunamis are impulsively generated waves that can be very destructive. They are generated by move-ment of the sea floor due to faulting, submarine landslides, or volcanic action. They have wavelengths of 130 to 165 kilometers and travel with velocities of 650 kilometers/hour (400 miles/hour). Their deep-water height is less than 3 meters, and they go unnoticed in the open ocean. However, as they approach the shoreline, their wavelength decreases and their height increases to many tens of meters. They have a very definite signature on tide-gauge records. Figure 10-7 shows the tide-gauge record at various ports for the destructive tsunami of April 1, 1946, in the Pacific Ocean basin. It was generated by subduction of the Pacific Plate under the North American Plate in the Aleutian Trench. The wave sped across the Pacific Ocean and did enormous damage in the Hawaiian Islands and coastal Alaska. Note the sudden "spikes" in the otherwise rather smooth sinusoidal shape of the tide curve.

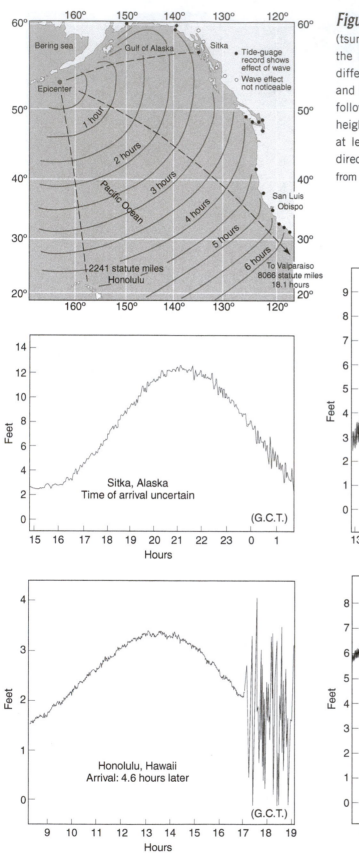

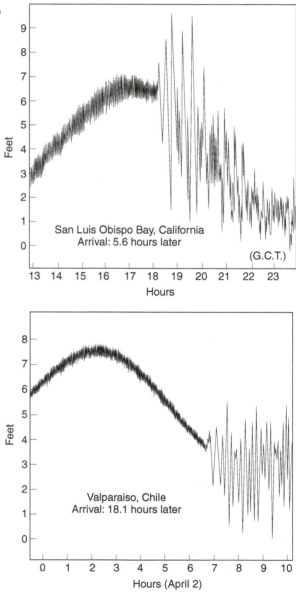

Figure 10-7 Records for a seismic sea wave (tsunami) of April 1, 1946, at selected points around the Pacific Ocean. Note that the tsunami arrived at different places at different stages of the tidal cycle, and that the first sign of its approach was a small rise followed by a larger fall in water level. The maximum height was not reached until the third or fourth crest, at least half an hour later. The map summarizes the direction of propagation and rate of travel. [Modified from C. K. Green, *Trans. Amer. Geophysical Union*, 1946.]

DEFINITIONS

Datum. The reference level to which tide levels are compared. The datum planes commonly used are mean low water or mean lower low water, which are the average levels of low tides taken over a 19-year period. These are also the datum planes ("0" ft or "0" m) used in constructing bathymetric charts.

Diurnal tides. Tides occurring once daily, one high and one low tide per tidal day.

Mixed tides. Complex tide curve, usually with two highs and lows of unequal height per tidal day.

Neap tides. The tides of lowest range, occurring twice monthly when the moon is a quadrature (so that the sun and moon are 90° apart).

Semidiurnal tides. Tides occurring twice daily. There are two high and two low tides per tidal day.

Spring tides. The tides of highest range, occurring twice monthly when the lunar and solar tides are in phase.

Storm surge. A rise above normal water level on an open coast due to strong winds blowing on-shore. Storm surge resulting from a hurricane or other intense storm also includes the rise in level due to atmospheric pressure reduction as well as that due to the winds. A storm surge is most severe when it occurs in conjunction with a high tide.

Tidal day (or lunar day). The time between two successive transits or passings of the moon over a local meridian. It is derived from the rotation of the earth relative to the movement of the moon about the earth. As the earth rotates once on its axis (24 hours) the moon has advanced in its orbit about the earth about 50 minutes; therefore the tidal day is 24 hours and 50 minutes long.

Tidal range. The difference in height between successive high- and low-tide levels.

Wind setup. The vertical rise in the water level on the leeward, or downwind, side of a body of water due to strong winds. Wind setup is similar to storm surge but the term is usually applied to reservoirs and smaller bodies of water.

Exercise *10*

*T*ides

NAME _____

DATE _____

INSTRUCTOR _____

1. Plot the tide heights at the proper time and day from the harbor tide record on the graph on the following page. Heights in the record are all measured above mean lower low water (MLLW) in the harbor. Connect the points with a straight line to produce a tide curve. Although straight lines do not produce the smooth curves like those shown in Figure 10-2, the record shows tide types and changes.

Harbor Tide Record

Day	Time (24 hour)	Height (meters)	Day	Time (24 hour)	Height (meters)
1	2400 (0000)	1.2	3	0955	1.8
	1130	2.1		1625	1.4
	2200	1.3		2205	1.7
2	0430	2.0	4	0720	1.4
	1110	1.2		1345	2.1
	1815	1.7		2205	1.6
3	0320	1.3		2345	1.8

(a) What type of tide is shown on day 1 _____, day 2 _____, day 3 _____, and day 4 _____?

(b) The least tidal range in the set is _____ meters on day _____.

(c) What is the elevation of the mean high water for the 4 days? _____ meters

(d) The datum for your curves is mean lower low water (MLLW). You have a small sailboat that draws 1.7 meters (5 $\frac{1}{2}$ feet), and you wish to sail it through a passage that is underlain by a rocky reef that is just exposed at MLLW. On what days and at what time would you sail? _____

_____

_____

_____

_____

_____

_____

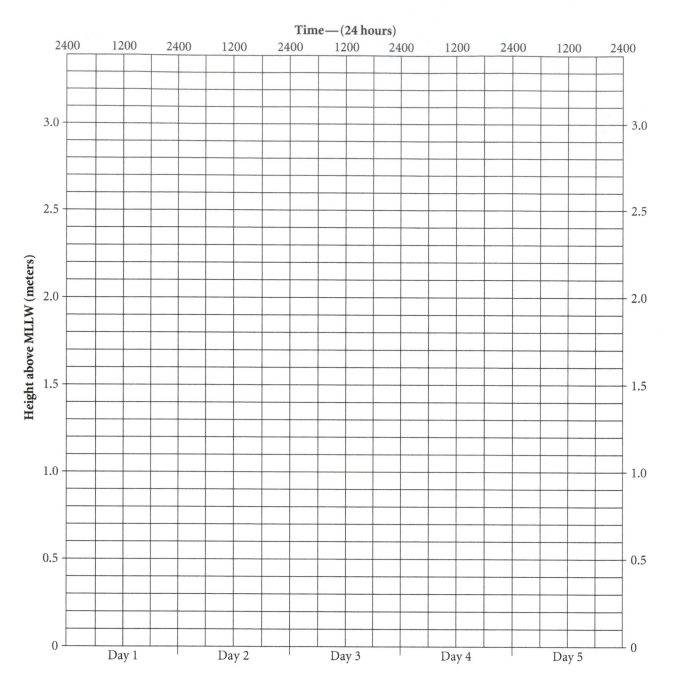

2. Just before midnight on day 5, the harbormaster noticed some unusual events in the harbor. A more detailed tide record was obtained, which is shown below. Add these points to day 5 on the tide curve of Question 1.

    (a) What event is suggested by these supplemental data, and what causes these events to occur?

_____

_____

_____

_____

Day	Time (24 hour)	Height (meters)	Notes
4	2345	1.8	Last regularly scheduled observation
5	0000	1.7	Sudden fall of water in harbor
	0020	3.1	Rapid rise of water
	0035	0.6	
	0048	3.4	Upper limit of tide gauge
	0100	0.2	Tide gauge fails — site abandoned

    (b) Which part of the event, the trough or the crest, entered the harbor first? _____

    (c) Is this event unusual? If not, how might knowledge of this be of survival value?

_____

_____

_____

_____

_____

3. (a) Using the travel time diagram in Figure 10-7, suggest the time required for a tsunami generated in the Aleutian Trench to arrive at the following locations:

    Sitka, Alaska _____

    Honolulu, Hawaii _____

    San Luis Obispo, California _____

    Valparaiso, Chile _____

    (b) What is the average velocity of the tsunami between the Aleutian Trench and Hawaii, 3600 kilometers distant? _____ kilometers/hour.

Between the Aleutian Trench and Valparaiso, Chile? _____ kilometers/hour.

Give a possible explanation for this difference in velocity between the two places.

_____

_____

_____

    (c) Estimate the greatest height of the tsunami above the tide level when the tsunami arrived at:

    Sitka _____ meters

    Honolulu _____ meters

    San Luis Obispo _____ meters

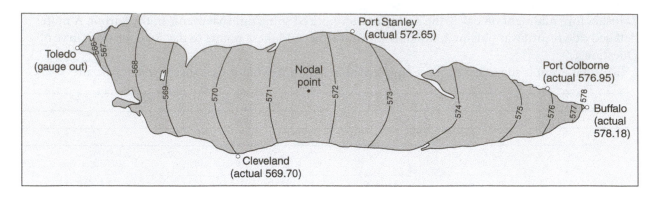

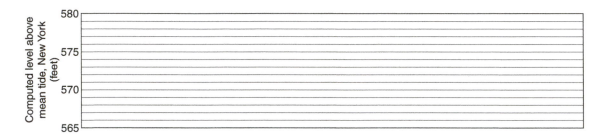

Figure 10-8 Effects of wind on surface-water level of Lake Erie. The contours show the water-level computer at 11:00 P.M. on November 8, 1957. [After I. A. Hunt, Jr., *Winds, Wind Set-up and Seiches on Lake Erie.* U.S. Lake Survey, U.S. Army Corps of Engineers, 1959.]

(d) What was the first evidence of the arrival of the tsunami on the shoreline at Honolulu — was it the trough or crest of the tsunami? _____

What crest in the series did the greatest damage? _____

4. Figure 10-8 shows the contours of water levels that were computed on Lake Erie during a storm on November 8, 1957.

(a) In the grid at the bottom of the figure, draw a profile of the computed water levels from Toledo (gauge out of water) to Buffalo.

(b) The nodal point is the point of no vertical change and thus represents the mean level of the lake. What is the value of the wind setup at Buffalo? _____

What is the maximum difference in water level between Buffalo and Toledo? _____

5. In the waters off southern California a small smeltlike fish, *Leuresthes tenuis* (the grunion), exhibits an interesting reproductive strategy finely timed to the tidal cycle. During spring tides from April to August, grunions come ashore shortly after the highest tides (which occur at night), and the female deposits eggs a few inches deep in the damp beach sand. The eggs are then fertilized by male grunions and are ready to hatch in 9–10 days, but only when the tidewater reaches them and they are agitated by surfaction. When will grunion eggs deposited at full moon on July 3 have their first opportunity to hatch?_____

Waves at Sea

OBJECTIVES:

■ To understand the generation and motion of wind waves at sea.

■ To understand the physics behind the limits of the height and velocity of deep-water wind waves.

■ To understand why the U.S. coastal states are subject to storm waves from some storm centers and not from others.

Waves at sea are created by winds blowing across the water surface and transferring energy to the water by the impact of the air. Small ripples develop first, and frictional drag on their windward side causes them to grow larger, or to collapse and contribute part of their expended energy to larger waves. Consequently, large waves capture increasing amounts of energy and continue to develop as long as the wind maintains sufficient strength and a constant direction. Generally, high winds of long duration produce large waves with long **wavelengths** and **wave periods.** As more and more energy is transferred to the water surface, waves become higher and longer, and travel with increasing **wave velocities.** A mariner's rule of thumb is that in an area of foul weather, **wave height** in feet will be approximately half the wind velocity in knots.* That is, winds of about 50 knots would develop waves about 25 feet high. Whether or not this approximation is reliable, the fact remains that 50-foot waves are not uncommon in the

---

*Recall that the knot is defined as 1 nautical mile per hour: 1 nautical mile is equivalent to 1.15 statute miles, or 1.85 kilometers.

open ocean, and waves more than 100 feet high have been reported. Our capability to predict waves at sea has obvious importance for navigation, transportation, and recreation.

Water Motion

It has long been observed that floating objects on the sea surface simply bob up and down or move with a slight rotary motion as waves pass underneath them. This is because water particles respond to the passing wave and move in circular orbits that decrease in diameter with depth (Figure 11-1). At a depth equal to about one-half the wavelength, the orbital diameters of the water particles are only $\frac{1}{25}$ of those at the surface, and, for all practical purposes, we may consider this level as the maximum depth of wave motion. In water deeper than half the wavelength, the orbiting particles do not contact the ocean bottom, whereas at depths shallower than half the wavelength the orbits are flattened by frictional resistance, they lose energy, and the wave is said to "feel the bottom." Geologists recognize this depth, which is called the **wave base,** as the maximum at which waves can move particles and erode fine sediment on the sea floor.

The Velocity of Deep-Water Waves

The velocity of deep-water waves (where water depth is greater than half the wavelength) is a function of their wavelengths; that is, the longer a wave, the faster it travels. However, the energy contained in a group of waves is transmitted at half

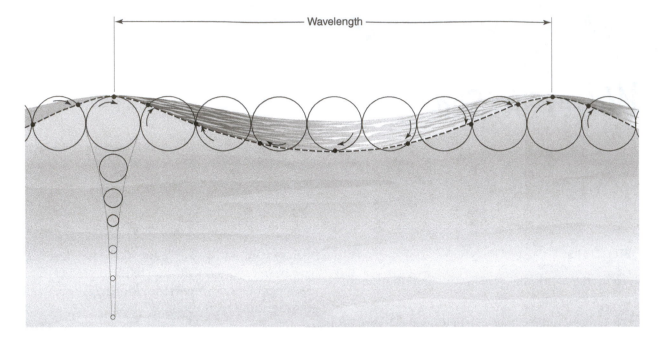

Figure 11-1 Cross section of an ocean wave traveling from left to right shows the wavelength as distance between successive crests. The time it takes for two crests to pass a point is the wave period. The circles are orbits of water particles in the wave. At the surface their diameter equals the wave height. At a depth of half the wavelength (left), the orbital diameter is only 4 percent of that at the surface. [After W. Bascom, "Ocean Waves." Copyright © 1959 by Scientific American, Inc. All rights reserved.]

the velocity of the individual wave. The reason is that waves at the front of a wave group decay and lose energy as they raise the water surface, and they are replaced by waves from behind (see Figure 11-2). It is somewhat analogous to a marching band in which the individual members walk at 5 kilometers per hour, but because the players in front leave the row to execute some other maneuver, the front row advances at only 2.5 kilometers per hour.

Fully Developed Sea

The development of waves in deep water is complex but may be attributed to three primary factors. These are the wind speed, the duration of wind, and the **fetch** of the wind (the distance of open surface across which the wind blows). In a discussion of wave development, the term **sea** refers to the occurrence on the sea surface within the fetch area of irregular waves of many periods coming from many directions. A **fully developed**

sea is formed when the speed of a given wind lasts long enough, and the wind has enough open water to work upon, to produce the maximum wave height that can be maintained by the wind. The necessary combination of sufficient duration and sufficient fetch rarely occurs for winds of high speed, but it is possible for most light winds. Table 11-1 shows the minimum fetch and duration required for various wind speeds to set up fully developed seas.

If conditions of duration and fetch permit a fully developed sea to form, we can predict the characteristics of the resulting waves as shown in Table 11-2. Since the data in this table are based on field observations, they are reasonably good estimates. The highest waves can be estimated statistically but cannot be predicted exactly. Therefore, for a 30-knot wind developing a full sea, we can expect that the highest wave in ten will be about 28 feet (8 meters) high, but we cannot predict when it will occur.

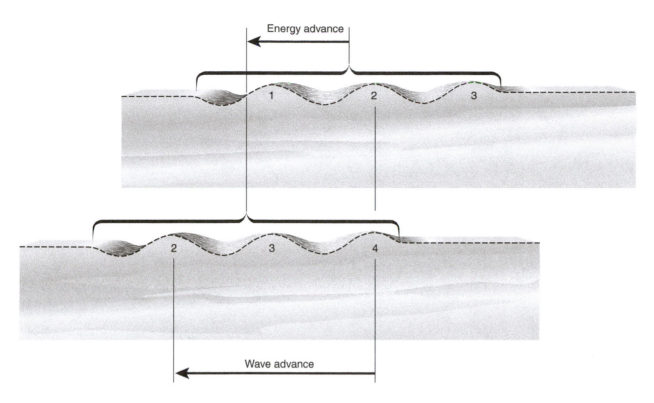

Figure 11-2 A moving train of waves advances at only half the speed of individual waves. At the top is a wave train in its first position. At the bottom, the train and its energy have moved only half as far as wave 2. Meanwhile wave 1 had died, but wave 4 has formed at the rear of the train to replace it. Waves arriving at shore are thus remote descendants of the waves originally generated. [After W. Bascom, "Ocean Waves." Copyright © 1959 by Scientific American, Inc. All rights reserved.]

The diagram in Figure 11-3 shows the characteristics of waves from different source areas that strike the California coast. Such diagrams are useful for planning purposes because they aid in predicting the kind of waves that would strike a given

TABLE 11-1

Minimum fetch and duration required for selected wind speeds to set up fully developed seas

Wind speed (knots)	Fetch (nautical miles)	Duration (hours)*
10	10	2
20	70	10
30	280	23
40	710	42
50	1420	69

*Duration times rounded off to the nearest hour.

part of the shoreline with a specific orientation, such as a south- or west-facing coast. Note that the waves with the longest period, and thus the ones containing the most energy, come from distant storms with large, unobstructed fetch areas. The waves that have traveled out of their source area are called **swell.** They are more regular, and have greater wavelengths and flatter crests, than the seas that generated them. Swell is usually denoted by source direction, such as *south* or *north swell.*

Table 11-3 shows the criteria for fully developed seas at differing wind speeds and the parameters of the waves that are produced. For example, for a wind speed of 30 knots, a fetch of 280 nautical miles and a storm duration of 23 hours are required for a fully developed sea. The resulting waves would have an average height of about 14 feet (4.1 meters) and a period of 8.6 seconds. The table shows that for storms that cover a large area, with only moderate subhurricane wind velocities, waves more than 30 feet high can develop. Thus

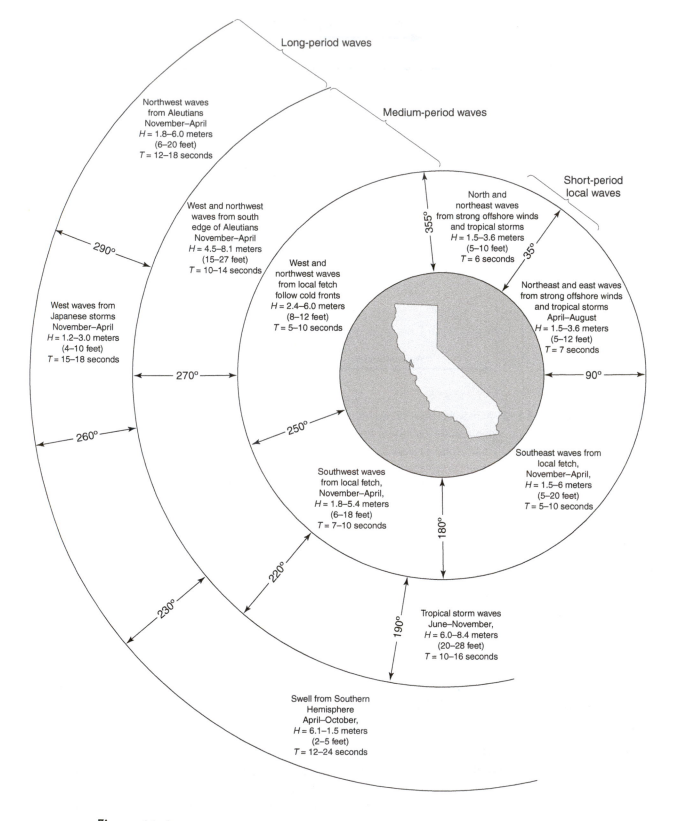

Long-period waves

Medium-period waves

Short-period local waves

Northwest waves
from Aleutians
November–April
H = 1.8–6.0 meters
(6–20 feet)
T = 12–18 seconds

West and northwest
waves from south
edge of Aleutians
November–April
H = 4.5–8.1 meters
(15–27 feet)
T = 10–14 seconds

West and
northwest waves
from local fetch
follow cold fronts
H = 2.4–6.0 meters
(8–12 feet)
T = 5–10 seconds

North and
northeast waves
from strong offshore winds
and tropical storms
H = 1.5–3.6 meters
(5–10 feet)
T = 6 seconds

Northeast and east waves
from strong offshore winds
and tropical storms
April–August
H = 1.5–3.6 meters
(5–12 feet)
T = 7 seconds

West waves from
Japanese storms
November–April
H = 1.2–3.0 meters
(4–10 feet)
T = 15–18 seconds

Southeast waves from
local fetch,
November–April,
H = 1.5–6 meters
(5–20 feet)
T = 5–10 seconds

Southwest waves
from local fetch,
November–April,
H = 1.8–5.4 meters
(6–18 feet)
T = 7–10 seconds

Tropical storm waves
June–November,
H = 6.0–8.4 meters
(20–28 feet)
T = 10–16 seconds

Swell from Southern
Hemisphere
April–October,
H = 6.1–1.5 meters
(2–5 feet)
T = 12–24 seconds

355°   35°   90°   180°   190°   220°   230°   250°   260°   270°   290°

Figure 11-3 Diagram for the California area showing the characteristics of waves from different directions. H refers to the wave height, T to the wave period. [After D. L. Inman, U. S. Army Corps of Engineers Beach Erosion Board Technical Memorandum.]

TABLE 11-2

Characteristics of waves resulting from selected wind speeds in a fully developed sea

Wind speed (knots)	Average height		Average length		Average period (seconds)	Highest 10 percent of waves	
	(feet)	(meters)	(feet)	(meters)		(feet)	(meters)
10	.9	.27	28.0	8.5	2.9	1.8	.55
20	5.0	1.5	111.0	33.8	5.7	10.2	3.1
30	13.6	4.1	251.0	76.5	8.6	27.6	8.4
40	27.9	8.5	446.0	135.9	11.4	56.6	17.2
50	48.7	14.8	696.0	212.2	14.3	98.9	30.2

fetch and storm duration are almost as important as wind velocity in producing large waves.

DEFINITIONS

Fetch. The length of unobstructed open sea surface across which the wind can generate waves.

Fully developed sea. The waves that form when wind blows for a sufficient period of time across the open ocean. The waves of a fully developed sea have the maximum height possible for a given wind speed, fetch, and duration of wind.

Sea. Local irregular waves of many periods and from many directions. A sea forms within storm areas or when local winds are blowing over the sea surface.

TABLE 11-3

Conditions necessary for a fully developed sea at given wind speeds and the parameters of the resulting waves

Wind speed (knots)	Fetch (nautical miles)	Duration (hours)	Average height		Average length		Average period (seconds)
			(feet)	(meters)	(feet)	(meters)	
10	10	2	.9	.27	28	8.5	3.0
12	18	4	1.4	.43	40	12.2	3.4
14	28	5	2.0	.61	55	16.8	4.0
16	40	7	2.8	.85	71	21.6	4.6
18	55	8	3.8	1.2	90	27.4	5.0
20	75	10	4.9	1.5	111	33.8	5.7
22	100	12	6.3	1.9	135	41.2	6.3
24	130	14	7.8	2.4	160	48.8	7.0
26	180	17	9.5	2.9	188	57.3	7.4
28	230	20	11.4	3.5	218	66.4	8.0
30	280	23	13.6	4.1	251	76.5	8.6
32	340	27	16.0	4.9	285	86.9	9.0
34	420	30	18.6	5.7	322	98.2	9.7
36	500	34	21.4	6.5	361	110.1	10.3
38	600	38	24.5	7.5	402	122.6	10.9
40	710	42	27.9	8.5	446	136.0	11.4
42	830	47	31.5	9.6	491	149.7	12.0
44	960	52	35.4	10.8	540	164.6	12.6

Swell. Waves that have traveled a long distance from the generating area and have been sorted out by travel into long waves of the same approximate period.

Wave base. The plane or depth to which waves may erode the bottom in shallow water.

Wave height. The difference in elevation between the crest and trough of a wave.

Wave period. The length of time, in seconds, required for a wave to pass a fixed point.

Wave velocity. The wavelength divided by the wave period (in feet per second or meters per second).

Wavelength. The distance, in feet or meters, between equivalent points (crests or troughs) on waves.

Exercise 11

Waves at Sea

1. The distance from San Pedro, California, to Avalon, on Santa Catalina Island, is about 25 nautical miles (nm). The island is almost due south (180°) from the Los Angeles Harbor at San Pedro.

    (a) What is the minimum wind speed that can set up a fully developed sea in this channel with a north wind? Refer to Table 11-1 and interpolate between fetch distances 10 nm and 70 nm (rounded to nearest whole number)._____knots, _____ kilometers/hour.

    (b) How long must the north wind blow for a fully developed sea to develop (interpolation required)? _____ hours.

    (c) By inspection of Table 11-2 or Table 11-3, using the fully developed sea wind speed you determined in (a), would the resulting wave heights be a problem to boaters in the Catalina Channel? _____ If not, why not? _____

    (d) When the fetch is 280 nm (or more), as it is in the winter in this area, what would be the wave height of a fully developed sea? _____ meters _____ feet. Would this wave height be more or less of a problem for boaters than the wave heights determined in the previous question? _____
_____

2. In Florida many summer storms come from the southeast with an enormous fetch. If the average duration of these storms is about 2 days, what is the wind speed that would set up a fully developed sea (Table 11-1)? _____ knots.

    What would be the resulting wave height (Table 11-2 or 11-3)? _____ meters _____ feet.

3. Figure 11-3 was developed by scientists at Scripps Institute of Oceanography for the U.S. Army Corps of Engineers, the government agency that is responsible for shoreline protection and harbor improvement for the entire country, including the Great Lakes. The diagram has obvious uses for shoreline engineering.

(a) From what sector do the largest waves strike California during the winter? _____

(b) From what sector does the largest swell come in the summertime? _____

What is the range of potential periods and heights? _____ seconds _____ meters

(c) Waves of what period (long, medium, short) produce the highest waves?

4. (a) What is the group velocity of storm waves with an individual wave velocity of 100 kilometers per hour? _____ kilometers/hour.

(b) How long would it take such waves to reach the New Jersey shoreline from a storm center in the Atlantic Ocean 1500 kilometers away? _____ hours.

(c) Why are long waves the first to arrive at the coast from an open-ocean distant storm? _____

_____

_____

_____

_____

_____

_____

_____

_____

_____

Waves in Shallow Water and Beach Erosion

OBJECTIVES:

■ To understand the characteristics of wind waves as they move from deep water to shallow water.

■ To discover what causes waves to peak and break as surf and how they generate longshore currents and dangerous rip currents.

■ To investigate the dynamics of sand, beach drifting, and beach erosion.

In Exercise 11 we discussed the origin and nature of deep-water waves in the open ocean. When these waves move into shallow water and strike the shoreline a transformation takes place, producing breakers, or surf, which can create or destroy beaches. Whereas most of us regard a beach as a place of recreation where we can relax, watch the surf, and get a tan, geologists and engineers recognize beaches as the last rampart of protection of the land against the sea. If we do not manage this resource wisely, the contest between human ingenuity and the ocean's relentless pounding will be won by the sea. Beaches along both coasts of the United States have been deteriorating badly in the course of the past few decades, and in some places they have completely disappeared. Where this occurs, valuable recreational land is lost and coastal-zone property endangered.

Shallow-Water Waves

Shallow-water waves are defined as those that are traveling in water whose depth is half the wavelength or less. Recall that the water motion in a deep-water wave is circular and that the diameter of the orbits decreases downward until, at a depth equal to about half the wavelength, water motion ceases. However, when a wave moves into shallow water, a drastic change takes place. The orbits of water particles at depth become flattened, and those in contact with the sea floor simply move back and forth (Figure 12-1). The wave is said to "feel the bottom" and, as a result, the wave velocity and length decrease, and the wave form steepens until it becomes unstable and spills against the shoreline as breakers, or surf. At this point the oscillations of the water particles cease and the water motion is all in one direction, toward the beach. Figure 12-2 shows this transformation from swell to shallow-water wave to surf. Whether a wave is classified as a shallow-water one or not depends on both the wave itself and the basin within which it is traveling. Figure 12-3 shows the relationship between wave period and water depth. Because the length of a wave is proportional to the square of its period, the diagram also shows the relationship between wavelength and depth of water. We see that very short-period waves, such as ripples, are deep-water waves even in relatively shallow water. Sea and swell are also deep-water waves in water from about 100 to 1000 feet deep. However, in water shallower than 100 feet, sea and swell become shallow-water waves. In the deep ocean, with an average depth of about 15,000 feet, all waves with periods greater than 80 seconds are shallow-water waves. In this category we find the tides and tsunamis, or so-called tidal waves. Tsunamis are long-period sea waves produced by submarine earthquakes, volcanic eruptions, or landslides.

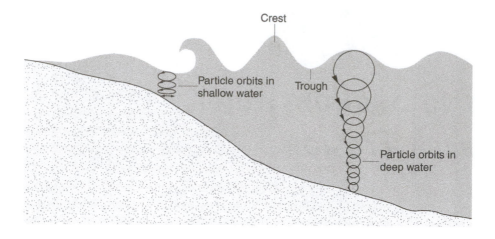

Figure 12-1 When waves approach a shallow bottom near the coast, they slow and steepen, and the circular orbits flatten and become smaller. The diagram shows considerable vertical displacement. [After R. A. Davis, Jr., *The Evolving Coast,* Scientific American Library, W. H. Freeman and Company. Copyright © 1997.]

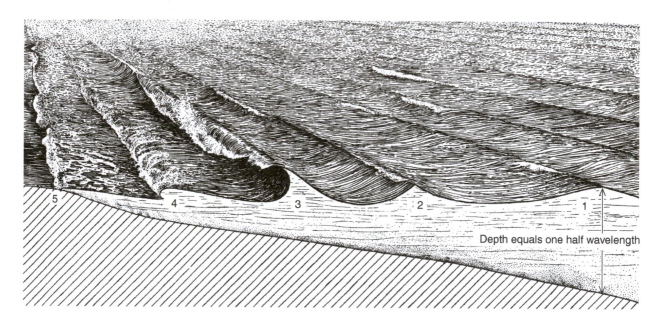

Figure 12-2 (1) A wave breaks up at the beach when swell moves into water shallower than half the wavelength. (2) The shallow bottom raises wave height and decreases length. (3) At a water depth of 1.3 times the wave height, water supply is reduced, and the particles of water in the crest have no room to complete their cycles; the wave forms and breaks. (4) A foam line forms and water particles, instead of just the wave form, move forward. (5) The low remaining wave runs up on the beach face as "swash," or uprush. [After W. Bascom, "Ocean Waves." Copyright © 1959 by Scientific American, Inc. All rights reserved.]

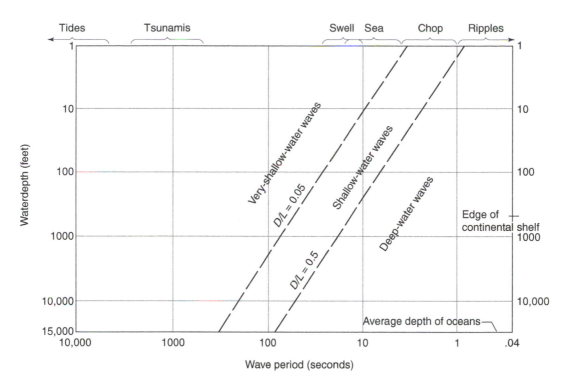

Figure 12-3 Wave characteristics. [After W. Bascom, *Waves and Beaches*. Doubleday and Company, New York, 1964.]

They may travel unnoticed across the ocean for thousands of miles from their point of origin and build up to great heights over shallow water.

Table 12-1 shows the relationship between selected wave periods, the calculated wavelength and wave velocity, and the water depth at which the wave feels the bottom or becomes a shallow-water wave. Note that wind-generated waves with periods greater than 14 seconds are capable of moving sediment at depths as great as the edge of the continental shelf. Most wind waves have periods between 5 and 25 seconds.

TABLE 12-1

The relationships between selected wave periods, calculated wavelengths and velocities, and water depths at which wave feels bottom

Wave period (seconds)	Wavelength* (feet)	Approximate velocity (miles per hour)	Water depth† (feet)
6	184	21	92
8	326	28	163
10	512	35	256
12	738	42	369
14	1000	49	500
16	1310	56	655

*Length is equivalent to 5.12 times the wave period squared.
†Depth is equivalent to half the wavelength.

Wave Refraction

Shallow-water waves are subject to **refraction** over humps or depressions of the sea floor, and to **reflection** from seawalls or breakwaters. Refraction occurs when a wave moves into shallow water at some angle other than parallel to the shoreline. The part of the wave crest in the shallowest water is slowed the most, whereas the part of the wave in deeper water moves forward at a higher velocity. The result is a bending of the wave crest and a concentration or dissipation of energy at the shoreline. An example of a refraction pattern at sea is shown in Figure 12-4. We can determine the relative amount of concentration or dissipation of wave energy by drawing lines perpendicular to the wave crests, known as **orthogonals,** on a diffraction diagram. The wave's energy, or its ability to do work on a shoreline, is the same between orthogonals drawn at equally spaced intervals along the wave crest. By tracing the orthogonals shoreward on the crests of successive waves of selected periods or lengths, we can determine how wave energy is concentrated or dissipated at the **surf zone** (Figure 12-5). The **wave energy coefficient,** e, which is the relative amount of concentration or dissipation of energy, may be calculated by simply dividing the distance between two orthogonals at the shoreline by the distance between the same two orthogonals in deep water. When waves are focused by wave refraction on a point, as in Figure 12-5, the distance between adjacent wave orthogonals at the shore decreases and e becomes less than 1. In contrast, when wave energy is dissipated, as along the beach in Figure 12-5, the distance between adjacent wave orthogonals at the shore becomes greater, resulting in e greater than 1. Tracing wave orthogonals shoreward thus provides an estimate of the relative energy expended on equivalent lengths of the shoreline. Low-energy shorelines have $e > 1$ and

Figure 12-4 Wave refraction, Bluff Cove California. [© Bruce Perry/Department of Geological Sciences at CSU Long Beach]

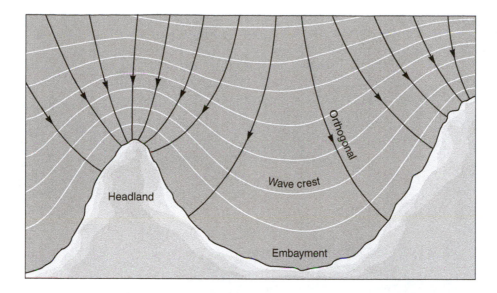

Figure 12-5 This wave refraction diagram shows how the energy of the wave front is concentrated by refraction around the small headland area. The same amount of energy enters the bay, but is spread at the beach over wide areas. The horizontal lines are wave fronts; the vertical lines (orthogonals) divide the energy into equal units for purposes of investigation. Such studies are vital preliminaries to design of shoreline structures. [From R. A. Davis, Jr., *The Evolving Coast*, Scientific American Library, W. H. Freeman and Company. Copyright © 1997.]

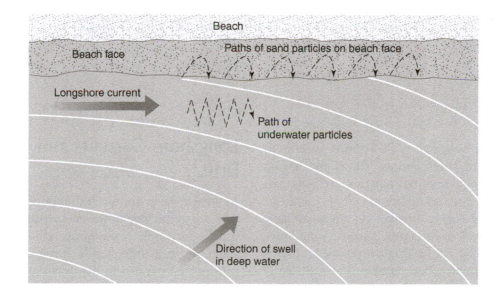

Figure 12-6 Longshore currents flowing parallel to the beach are formed when waves approach the beach face at an angle. Incoming waves carry sand grains up the beach face at an angle, but backwash from the wave returns the grains directly down the face to the surf zone. This alternating motion causes the zig-zag pattern seen in the figure, and results in movement of sand along the beach face by the wave-generated longshore current. [After W. Bascom, "Beaches." Copyright © 1960 by Scientific American, Inc. All rights reserved.]

Figure 12-7 Groins, which are dams of wood, stone, or concrete, are built perpendicular to a beach to trap sand. The groins in this photograph are on the Atlantic coast at Long Beach, Long Island. Photographed on 9/20/1994 [© John Wark/Airphoto]

are usually depositional, with a sandy beach. On the other hand, high-energy shorelines having $e < 1$ are erosional and may have a gravel (shingle) beach, or a sea cliff and no beach.

The height of a breaker is approximately equivalent to 0.78 of its depth. Thus we would predict that a breaker 7.8 feet high would occur in water 10 feet deep, provided that swell of sufficiently long period were approaching the shoreline. This type of approximation is important in determining the height that piers should be built above mean high water or some other datum. Obviously fishing or commercial activity from a pier 15 feet high would be impeded if the pier were exposed to 18-foot waves (water depth about 23 feet).

Longshore Currents and Littoral Drift

As a rule, waves approach the shoreline at an angle and are refracted; however, because refraction is usually incomplete, the waves strike the shore at a slight angle. Consequently some of the water is transported parallel to the beach and a weak **long shore current** (which flows parallel to the shore) is created (Figure 12-6). The current is like a river on land, and is capable of moving sand along the beach, a process known as *beach drifting*. The earth material moved along the beach by the longshore current is known as **littoral drift.** When an obstruction, such as a groin or jetty, is placed in the

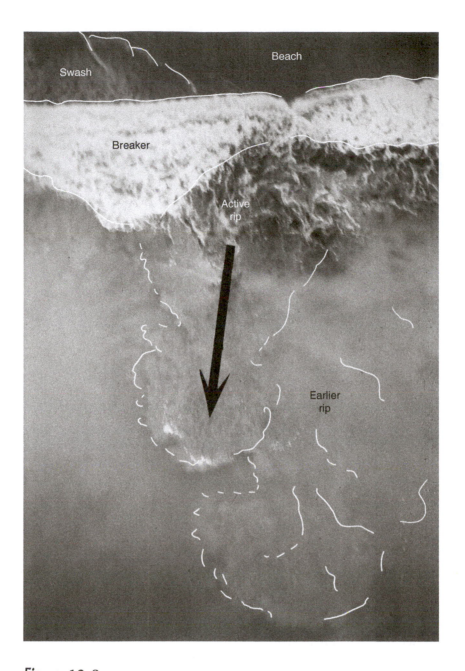

Figure 12-8 Rip currents in the breaker zone near Carpinteria, California, in February 1969. Two pulses of flow can be seen: an earlier jet is seaward and deeper then the active jet. Each pulse is probably the product of a series of breakers. The older pulse is the result of rip generated a few minutes earlier. [Donn Gorsline]

path of the current, a buildup, or accretion, of littoral drift results on the upstream side and erosion occurs on the downstream side (Figure 12-7). The extent of this accretion and erosion depends on the velocity and persistence of the current and the sup-

ply of sand. Also, like a river, the longshore stream may overflow. If we view the banks of the river as the beach and the surf zone, we realize that as water builds up in the longshore current, it must eventually "overflow" and return seaward. It does

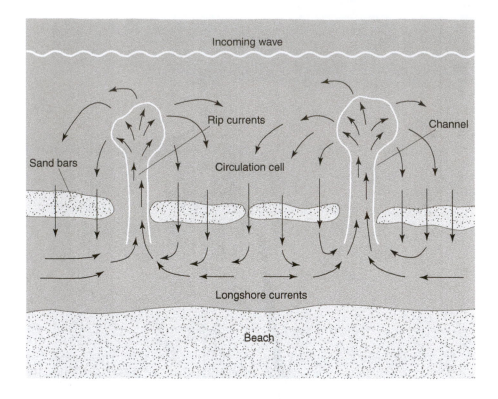

Figure 12-9 Diagram of the nearshore current system. [From R. A. Davis, Jr., *The Evolving Coast,* Scientific American Library, W. H. Freeman and Company. Copyright © 1997.]

so in the form of **rip currents,** improperly called rip tides, which can pose a formidable hazard to swimmers when the waves are high. They can be identified by the presence of a gap in the wave forms of the breakers, white foamy water extending seaward beyond the turbulence of the surf zone, and streaks of sediment or floating objects moving seaward in the rip (Figure 12-8). A diagram of the nearshore system of circulation and currents is shown in Figure 12-9.

The longshore current, unlike a river, may reverse its direction as a consequence of differing wave approaches, but it never stops flowing. Thus, where the supply of sand to a given point along the beach is less than the amount removed by longshore currents, erosion and loss of beach ensue. Although beach erosion may result from natural causes—such as drought, which decreases sand supply from rivers—coastal engineering works are more often directly responsible.

Structures of considerable height that extend seaward for some distance are effective barriers to littoral drift. But short groins are usually constructed for shoreline stabilization and only temporarily disrupt the longshore transport of sand (see Figures 12-10 and 12-11. Table 12-2 indicates the magnitude of the problem in southern California. Keep in mind that for every cubic yard of sand trapped or stored by a groin or breakwater, the beaches that are downcurrent are deprived of an equivalent amount, and the loss of beach width by erosion inevitably occurs. The same problem is also prevalent in the Great Lakes and along the East and Gulf coasts of the United States. The figures shown for accretion are even more striking when we consider that a cubic yard is roughly equivalent to 1 square foot of beach. That is, if 100,000 cubic yards of sand are lost, this volume is equivalent to beach retreat of about 1 foot along a length of 100 feet. It may be seen that if loss of

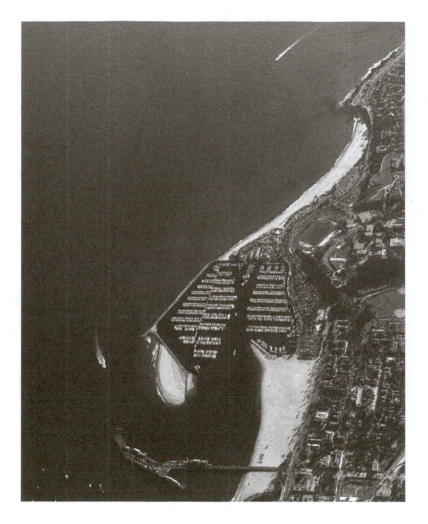

Figure 12-10 Jetty at Santa Barbara, California, causes sand transported by longshore currents to be deposited as a shoal downdrift, near the end of the breakwater. Sand must be periodically dredged from the harbor to beaches farther east (lower right of photograph) [© David Saffir photo/www.davidsaffir.com]

sand continues for many years it can seriously diminish a recreational area. Finally, repair of the damage is costly.

Barrier beaches, also known as offshore bars, ring the entire East and Gulf coasts of the United States from New York to the tip of Florida to Padre Island, Texas. Atlantic City, New Jersey, Kitty Hawk, North Carolina, and Miami Beach, Florida are all examples of barrier beaches that are subject to inundation and extreme erosion during tropical storms and hurricanes (see Exercise 13). Miami Beach spent $10 million in the 1980s to dredge sand from offshore and place it along approximately 10 miles of city beach front. This replenished beach was washed back into deeper water by winter storm waves in just 3 years. An outstanding example of extreme erosion and loss of habitat is Hog Island, Virginia (Figure 12-12). In the late 1800s it was the site of lavish fishing and hunting clubs and was covered by a pine forest. A hurricane in 1933 inundated the island and destroyed the forest and town, which is now under water hundreds of meters from shore.

The natural transport of longshore drift is shown dramatically in three photographs of Sand Beach, New Jersey, taken over a period of 23 years (Figure 12-13a, b, and c). Note the growth of the spit within that period of time and the direction of the longshore current.

Figure 12-11 Historical aerial photograph Santa Monica, California 1947. Image shot 1947.
[© Aerial Archives/Alamy]

TABLE 12-2

Rates of littoral drift and accretion at shoreline structures for selected southern California localities

Location	Accretion rate (10^3 cubic yards per year)
Santa Barbara breakwater (Figure 12-10)	280
Santa Monica breakwater (Figure 12-11)	259
Redondo Beach	30
Anaheim Bay jetties	175
Balboa Bay jetty	72

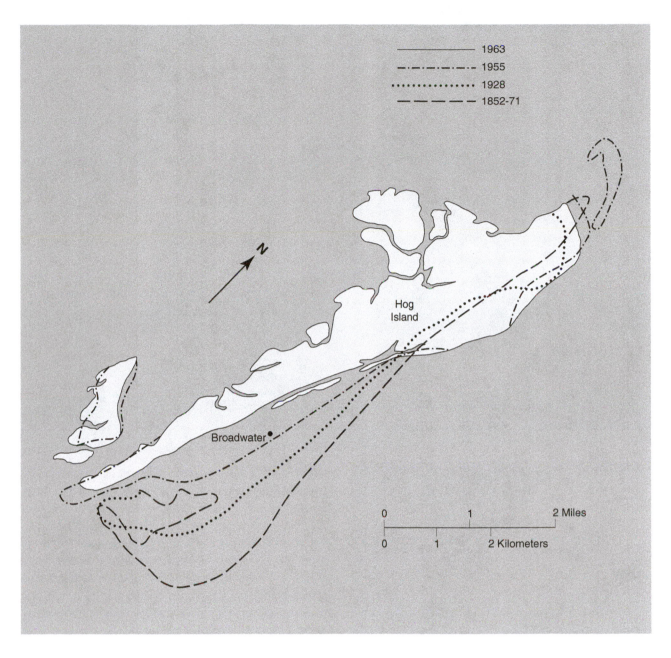

1963
1955
1928
1852-71

Hog
Island

Broadwater

0   1   2 Miles

0   1   2 Kilometers

Figure 12-12 Hog Island is a prime example of extreme erosion of barrier islands along the east coast of the United States. The main town of Broadwater had a population of about 300, 50 houses, a lighthouse, cemetery, and school. A hurricane in 1933 over-topped the island, destroying much of the town and killing the protective pine forest. By the 1940s all inhabitants had left, and the town is now under several meters of water.

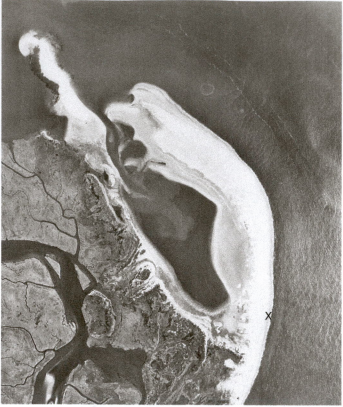

a

b

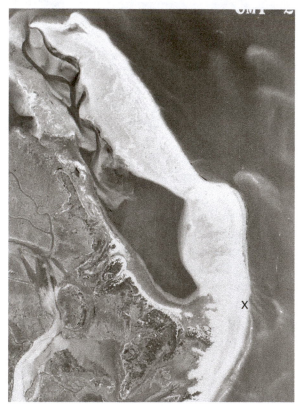

Figure 12-13 The natural transport of longshore drift over a 23-year period at Sandy Beach, New Jersey (designated by the cross in all photos). (a) 1940: The first stage in the development of a spit; the longshore current runs toward the north. (b) 1957: The spit extends with time in the direction of longshore transport of sediment. (c) 1963: The spit joins the land and the recurved spit grows around the tip. Also note that the main spit has screened the shore from wave action. [From C. S. Denny et al., *A Descriptive Catalog of Selected Aerial Photographs of Geologic Features in the United States.* U.S. Geological Survey Professional Paper 590, 1968]

c

DEFINITIONS

Littoral drift. The transport of sand, gravel, and other materials along the beach face by longshore (or littoral) currents.

Longshore (littoral) current. A current running parallel to the beach and generated by waves striking the shoreline at an angle.

Orthogonal. A line drawn perpendicular to wave crests so that refraction or bending can be visualized more clearly.

Reflection. The process by which the energy of wave is a returned seaward.

Refraction. The process by which the direction of a wave moving in shallow water at an angle to the bottom contours is changed. The part of the wave moving shoreward in shallower water travels more slowly than that portion in deeper water, causing the wave to turn or bend to become parallel to the contours.

Rip current. A current flowing seaward from the shore through gaps in the surf zone. The strength of the current is proportional to the height of the breakers striking the shoreline.

Surf zone. The nearshore zone along which the waves become breakers as they approach the shore.

Wave energy coefficient (e). Calculated as follows:

$$e = \frac{\text{width of orthogonals at shore}}{\text{width of orthogonals in deep water}}$$

Exercise 12

Waves in Shallow Water and Beach Erosion

NAME _____

DATE _____

INSTRUCTOR _____

Refer to the appropriate figures in the text of this exercise to answer questions 1–4.

1. What is the direction of the longshore current and beach drifting in Figure 12-7? Answer in terms of left and right sides of the photo. _____
_____

How did you determine this direction? _____
_____
_____

2. Examine Figure 12-10 and predict what will happen to Santa Barbara Harbor and the beaches downdrift if the harbor is not dredged and the sand moved downdrift of the pier. _____
_____
_____

3. Figure 12-14 depicts the nearshore oceanographic conditions at Santa Monica Beach, California (see also Figure 12-11). It should be noted that the pier is on widely spaced wooden pilings and has little influence on wave energy.
   (a) Why has sand built out behind the breakwater? _____
   _____
   _____

   (b) Indicate on the figure below the relative wave energy at points A–C. Use the terms *high*, *medium*, and *low*.

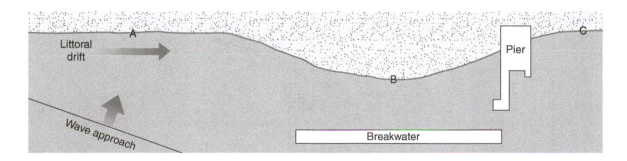

Figure 12-14 Diagram of a breakwater and beach at Santa Monica, California.

(c) Based upon the presence of the breakwater and wave direction shown in the figure, what condition of beach stability would you *predict* at point C: deposition, erosion, or no change? _____

_____

However, the beach appears stable without erosion or deposition. How might this be explained? _____

_____

_____

_____

4. Make a sketch of the wave refraction pattern around the small rocky island at Prouts Neck, Maine (Figure 12-4). Why does the beach extend farther seaward off the island than at adjacent areas?

5. Refer to the photos of Sandy Beach, New Jersey (Figure 12-13). (The right-hand side of the page is oriented directly north-south).
   (a) What is the direction of wave approach? _____
   (b) What is the direction of longshore drift? _____
   (c) If the scale of all three photos is identical (1 inch = 2250 feet), how much has Sandy Beach grown between 1940 and 1957? _____

   (d) Between 1957 and 1963? _____

6. Figure 12-15 shows wave-crest refraction shoreward. The dashed line at point Y represents a proposed breakwater, intended to complement the existing groin structure. Two wave orthogonals are sketched in for you.
   (a) Sketch in orthogonals beginning at points 1, 2, 3, 4, and 5.
   (b) Indicate along the shore in the diagram several places where you would expect the sand beach to become wider because of accretion, or to narrow because of erosion.

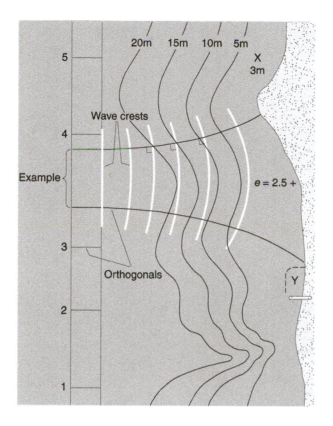

Figure 12-15 Wave refraction diagram for question 6. The numerals along the contour lines indicate water depths in meters. The wave energy coefficient, e, is 2.5 or more. Note the existing groin near point Y, and the dashed line indicating a proposed breakwater to be constructed next to the groin.

(c) What is the expected breaker height (in meters) at point X? _____

_____

(d) Where would the best board surfing be found (biggest waves)? _____

_____

(e) Where would the safest bathing beach be found? _____

_____

(f) Sketch in the predicted wave refraction pattern over the submarine canyon.
What would be the expected direction of longshore currents north of the canyon, that is, between the head of the canyon and the groin structure near point Y? _____

_____

_____

_____

(g) How would the longshore current affect the proposed breakwater at point Y? _____

_____

_____

_____

7. Refer to Figure 12-12, Hog Island. How much did the shoreline retreat between the mapping performed in 1871 and that of 1963? _____ miles, _____ kilometers, _____ miles/year.

*H*urricanes

OBJECTIVES:

■ To learn how and where hurricanes are spawned and how we measure their strength.

■ To understand the role of the oceans in the formation of hurricanes.

■ To appreciate the tremendous energy released in a hurricane and to respect the damage potential of hurricane winds, waves, and storm surge.

Tropical cyclones have a significant impact on coastal areas of the world. In the Atlantic Ocean they are called **hurricanes** (from *Hurican,* the Carib god of evil), whereas similar storms in the Pacific Ocean are called **typhoons.** They are a significant part of global weather systems because they transfer large quantities of water and warm, moist air from equatorial regions to higher latitudes. In the Caribbean Sea, the Gulf of Mexico, and the southeastern United States, hurricanes have shaped low-lying coastal areas and, at times, have caused dramatic losses of life and property. Even with the use of orbiting weather satellites, hurricane prediction and tracking remains a major challenge for meteorologists.

The Production of Hurricanes

The hurricane season extends from June 1 to October 1 in the North Atlantic Ocean. This is because hurricanes get most of their energy from the equatorial oceans in the summer and early fall, when seawater temperatures are warmest, usually above 25°C (about 80°F). Hurricanes begin as

tropical waves, areas of organized clouds 200–500 kilometers in diameter (Color Plate 2), in the eastern equatorial Atlantic. As hot air rises over the warm ocean water, the atmospheric pressure drops, and cyclonic (counterclockwise) circulation begins, with strong winds rushing toward the low-pressure center. At this point the wave becomes a *tropical depression.* It becomes a *tropical storm* when winds 65–118 kilometers/hour (39–74 miles/hour) develop. If winds exceed 118 kilometers/hour, the storm is called a hurricane (in the Atlantic and eastern Pacific oceans) or a typhoon (in the western Pacific Ocean).

Tropical storms and hurricanes have cyclonic circulation in the Northern Hemisphere; that is, they revolve counterclockwise. Air flows toward the center of the low-pressure cell, and the Coriolis effect causes it to be deflected to the right. Recall that Coriolis deflection is to the left in the Southern Hemisphere, so that cyclones in the southern tropics have clockwise rotation.

Naming Hurricanes

Prior to 1953, hurricanes were named for the year or month of their occurrence, for the writer who described their passage, or for places sustaining severe damage, like the Galveston Hurricane of 1900. During World War II, U.S. Navy and Army Air Corps meteorologists forecasting tropical cyclones in the Pacific Ocean began to informally give typhoons women's names, probably after their wives or girlfriends. Other naming schemes were women's names in alphabetical order, and by the international phonetic alphabet—Alpha, Bravo, Charlie,

| TABLE | 13-1 | | | | |

The Saffir-Simpson hurricane scale

Hurricane category	Maximum sustained wind speed		Storm surge (feet)	Minimum Surface pressure (mbar)	Damage
	kilometers/ hour	miles/ hour			
1	119–153	74–96	4–5	>980	Minimal
2	154–178	97–111	6–8	979–965	Moderate
3	179–210	112–131	9–12	964–945	Extensive
4	211–250	132–155	13–18	944–920	Extreme
5	>250	156+	18+	<920	Catastrophic

Delta, Echo, Foxtrot, and so on. In 1953 the U.S. Weather Bureau began to name Atlantic hurricanes with an alphabetical sequence of female names.

The current hurricane-naming method is one of alternating male and female names, such as the recent names Alberto, Beryl, Chris, Debby, and so forth. Many names have been retired to the hurricane "Hall of Fame" because their storms were so devastating—Andrew, Gilbert, and Camille, for example.

Measuring Hurricanes

The strength of hurricanes is measured by the **Saffir-Simpson scale,** which uses minimum sustained wind speeds to categorize storms from Category 1, a minimal hurricane, to Category 5, the most intense hurricane (Table 13-1). Hurricane Mitch, which devastated Central America, and Hurricane Georges, which struck the coastal United States in 1998, were Category 5 events.

Since 1944, aircraft have been used to study tropical disturbances that had the potential to develop into hurricanes. Wind speeds and atmospheric pressure changes are measured by technicians in aircraft flown by Air Force personnel, who are called "The Hurricane Hunters" (Figure 13-1).

Because hurricanes have diameters of hundreds of kilometers and may have dozens of convective storms embedded in them, they may affect coastal areas far from the main storm center. As they approach the shoreline, NOAA's Hurricane Warning Center in Miami Beach, Florida, may is-

sue a **hurricane watch,** a 36-hour advance notice of high winds and storm surge, or a **hurricane warning,** which means that hurricane winds and surge are expected within 24 hours.

Hurricane damage is caused by high winds, flooding, and storm surge. Surface winds can cause an abnormal rise in sea level, reaching from 1 to 6 meters above normal sea level (Figure 13-2). Along the Gulf Coast, the estimated 330 kilometer/hour winds of Hurricane Camille in 1969 caused a maximum **storm surge** of 8 meters, with 3-meter-high waves atop the surge. The end result was a 10-meter-high wall of water approaching the Mississippi coastline. Failure to evacuate caused significant loss of life. One survivor of a "hurricane party" was found floating on a sofa 8 kilometers further inland from the coastal apartment in which the party was held.

Not all tropical storms develop into hurricanes, and not all hurricanes develop damaging storm surge. Hurricanes tend to be cyclic, with periods during which relatively weak tropical storms approach the coastline followed by periods of stronger hurricanes. This is a dangerous sequence, because coastal residents tend to remember only the most recent tropical storm events. Cyclicity in hurricane development and strength may be related to El Niño/Southern Oscillation events (see Exercise 17), as severe El Niño events may suppress tropical storms in the Atlantic Ocean, while permitting stronger storms to approach the west coast of North America.

The relative strength of hurricanes can be assessed from the atmospheric pressure of each storm, measured in millibars (mb) or millimeters

Figure 13-1 Inside the eye of Hurricane Ike, a WC-130J Hurricane Hunter aircraft provides data from several different instruments including sensors on the aircraft, dropsondes and the new Stepped-Frequency Microwave Radiometer, called a Smurf. The Smurf provides surface-level winds giving forecasters at the National Hurricane Center yet another source of information. [U.S.Air Force Photo/Tech. Sgt. James B. Pritchett]

The Movement of Hurricanes

The advance of a tropical storm or hurricane is controlled by the prevailing surface currents of the equatorial Atlantic Ocean. Tropical storms are moved westward by the North Equatorial Current, but there is extreme variability in the latitudinal (north–south) location of the storm track. Tropical cyclones may move through the Caribbean Sea between South America and the east–west chain of islands east of Cuba and Hispaniola, or they may be deflected north of Cuba into the Bahamas. Storms approaching the Atlantic coast of North America may be deflected northward by the Gulf Stream, or their advance may be sufficiently rapid to move them onshore into Florida or Georgia. Tropical cyclones developing in the Caribbean may move westward to Central America, or as did Hurricane Mitch in 1998, they may be forced northward across the Gulf of Mexico by the Loop Current. Since 1949, seven Atlantic hurricanes have crossed into the eastern Pacific Ocean, becoming Northeast Pacific hurricanes or tropical storms.

Another factor that may influence the advance of tropical storms in the Atlantic Ocean is the presence of warm or cold weather fronts or cells on the North American continent during the approach of the storm. Strong clockwise (anti-cyclonic) winds of high-pressure weather cells may slow the approach of tropical storms, or even stop their approach entirely. Hurricanes "stalled" in the Gulf of Mexico or the Atlantic Ocean may intensify, degenerate, or be deflected to the northeast without making any landfall on the North American continent. The unpredictable nature of hurricane movement is perhaps best shown by the "zig-zag" approach of Hurricane Elena in September 1985 (Figure 13-3).

Forecasting Hurricanes

Because it is so critical to assess the strength of tropical cyclonic storms, the National Oceanic and Atmospheric Administration (NOAA) operates the National Hurricane Center in Miami, Florida. The National Hurricane Center distributes tropical weather, hurricane, and severe storm information to the public by television, radio, and the Internet. Its web site, at **<http://www.nhc.noaa.gov>**, con-

of mercury (mm Hg). Inches of mercury is used in aviation and in some weather reporting formats (see Table 13-3 for conversions). Minimum surface pressure, however, is not necessarily a measure of the destructive potential of a hurricane. The lowest atmospheric pressure ever measured was during Hurricane Gilbert in 1998 in the Caribbean Sea: 888 mb, which is a record low. However, Gilbert was not as destructive as some other less intense hurricanes because its **landfall** was onto rising terrain and into less densely populated areas of the United States and Mexico. Table 13-2 shows the 20 most intense North American hurricanes since 1900 based on atmospheric pressure in the storm cell. Table 13-3 lists the most expensive American hurricanes since 1900. For each storm, the damage estimates have been adjusted to 1996 dollars.

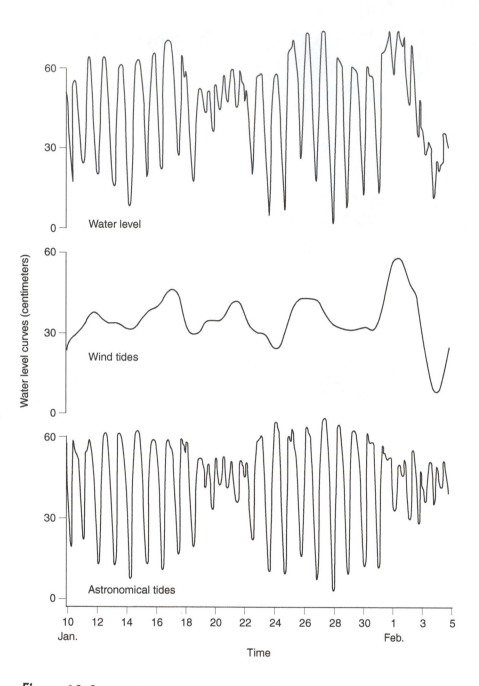

Figure 13-2 A tidal gauge on Mustang Island, Texas, recorded the water levels shown on the top plot. The bottom two curves show the contribution of storm surges (wind tides) and astronomical (sun and moon) tides to the total level of the upper curve. Note the extreme high water around February 1 due mainly to storm surge.

tains the latest forecasts for tropical storm activity, reconnaissance data, and historical storm data for the North Atlantic, Eastern Pacific, Central Pacific, Western Pacific, and other regions. Another "offi-

cial" source of information is the NOAA National Climatic Data Center, which contains information on hurricanes, El Niño, global climate, and other weather data. The NCDC has images and movies

TABLE 13-2

Most intense American hurricanes, 1900–1996

Hurricane	Year	Category	Pressure* (mb)	(in Hg)	(mm Hg)
Gilbert	1988	5	888	26.2	666.1
Florida Keys	1935	5	892	26.35	669.2
Camille	1969	5	909	26.84	681.7
Andrew	1992	4	922	27.23	691.6
Florida Keys/S. Texas	1919	4	927	27.37	695.2
Lake Okeechobee, FL	1928	4	929	27.43	696.7
Donna	1960	4	930	27.46	697.5
Galveston, TX	1900	4	931	27.49	698.2
Grand Isle, LA	1909	4	931	27.49	698.2
New Orleans, LA	1915	4	931	27.49	698.2
Carla	1961	4	931	27.49	698.2
Hugo	1989	4	934	27.58	700.5
Miami, FL/Pensacola, FL	1926	4	935	27.61	701.3
Hazel	1954	4†	938	27.70	703.6
SE FL/SE MS/AL	1947	4	940	27.76	705.1
North Texas	1932	4	941	27.79	705.9
Opal	1995	3†	942	27.82	706.6
Frederic	1979	3	946	27.94	709.7
Betsy	1965	3	948	27.99	710.9
Fran	1996	3	954	28.17	714.8

*Pressure exerted by the atmosphere at sea level is equal to 1 atmosphere, which is equal to 760 millimeters of mercury (760 mmHg), 1.01325 bars, 1013.25 millibars (mb), 29.92 inches of mercury (29.92 in Hg), and 14.7 pounds per square inch (14.7 psi)
†Hurricanes traveling more than 50 kilometers/hour.

TABLE 13-3

Most expensive American hurricanes, 1900–1996

	Hurricane	Year	Category	Damage estimate	Damage (in 1996 dollars)
1	Andrew	1992	4	$26.5 billion	$30.8 billion
2	Hugo	1989	4	$7.0 billion	$8.5 billion
3	Fran	1996	3	$3.2 billion	$3.2 billion
4	Opal	1995	3*	$3.0 billion	$3.1 billion
5	Frederic	1979	3	$2.3 billion	$4.3 billion
6	Agnes	1972	1	$2.1 billion	$7.5 billion
7	Alicia	1983	3	$2.0 billion	$3.0 billion
8	Bob	1991	2	$1.5 billion	$1.7 billion
9	Juan	1985	1	$1.5 billion	$2.1 billion
10	Camille	1969	5	$1.4 billion	$6.1 billion
11	Betsy	1965	3	$1.4 billion	$7.4 billion
12	Elena	1985	3	$1.2 billion	$1.8 billion
13	Gloria	1985	3*	$900 million	$1.3 billion
14	Diane	1955	1	$831 million	$4.8 billion
15	Erin	1995	2	$700 million	$718 million

*Hurricanes moving more than 30 miles per hour

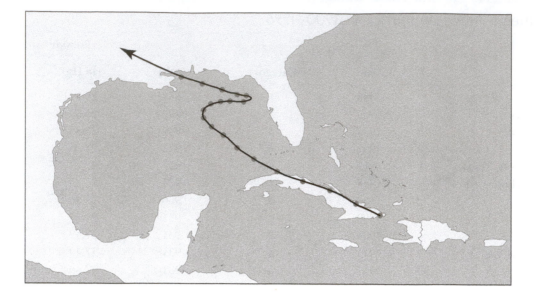

Figure 13-3 Perhaps the most remarkable hurricane approach in the last quarter-century was the track of Hurricane Elena in the Gulf of Mexico during September, 1995. Elena strengthened to hurricane status in the northern Caribbean Sea between Cuba and Hispaniola, then moved westward until it encountered the Loop Current in the Gulf of Mexico. The Loop Current deflected the storm to the northeast, where it "stalled" off Florida, then made a "retrograde" motion nearly due westward across the coastal areas of the Florida Panhandle and Alabama—before making landfall on the Mississippi coast.

of hurricanes, tropical storms, typhoons, and other severe storms at **<http://www.nodc.noaa.gov/ol/satellite/olimages.html>**. Other web sites providing hurricane information and graphics are listed in the Bibliography.

DEFINITIONS

Cyclonic circulation. Wind circulation around a low-pressure weather cell. In the Northern Hemisphere, when viewed from above, cyclonic winds follow a counter-clockwise rotation pattern.

Eye. The center of circulation of a hurricane or tropical cyclone, marked by calm wind speeds inside the center of the storm.

Hurricane. A tropical storm in the Atlantic Ocean, with cyclonic circulation and wind speeds greater than 118 kilometers/hour (74 miles/hour).

Hurricane Warning. Indicates that hurricane winds are expected in a designated coastal area within 24 hours.

Hurricane Watch. Indicates that hurricane winds are expected in a designated coastal area within 36 hours.

Landfall. The point where the eye of a tropical storm moves inland at a coastline.

Saffir-Simpson Scale. The standard for measuring hurricane strength, based on minimum *sustained* wind speeds.

Storm surge. Sea level elevation above normal levels at a coastline, caused by tropical storm winds "piling up" water as the storm approaches land.

Typhoon. A tropical storm in the Pacific Ocean, with cyclonic circulation and wind speeds greater than 118 kilometers/hour (74 miles/hour).

Exercise 13

Hurricanes

NAME _____

DATE _____

INSTRUCTOR _____

1. Use the information in Tables 13-2 and 13-3 to compare hurricane strength (storm category and atmospheric pressure) to hurricane damage.

    (a) Is there a relationship between strength and damage? _____
    (b) Does a Category 5 storm always produce more damage than one of Category 4? _____
    (c) Is there a predictable relationship between storm damage and meteorologically measurable data (wind speed, storm category, atmospheric pressure)? _____
    (d) What factors could produce more damage from a "weaker" hurricane? _____

2. Figure 13-4 is a hurricane tracking chart, modified from one available from the National Oceanic and Atmospheric Administration (NOAA). Using the information in Table 13-4, plot the location of the eye of this hurricane from July 12 to 14.

TABLE	13-4

Hurricane Zeke — July 1998

Date	Time	Latitude (°N)	Longitude (°W)	Wind speed miles/hour	kilometers/hour
7/12	0800	21.3	67.5	90	150
	2000	22.1	69.3	100	166
7/13	0800	23.0	71.5	115	192
	2000	24.2	73.8	125	208
7/14	0800	25.0	75.8	130	217
	1200	25.2	76.4	135	225
	1600	25.8	77.3	130	217
	2000	26.1	78.2	125	208

(a) If the radius of maximum wind from the eye of the hurricane is 150 nautical miles, and the radius of *all* storm winds is 300 nautical miles, predict the location and estimated time of landfall for Hurricane Zeke. Recall that one degree of latitude equals 60 nautical miles. _____
(b) At what date and time would you issue a hurricane watch for the Miami region? _____
(c) When would you issue a hurricane warning for the Miami region? _____

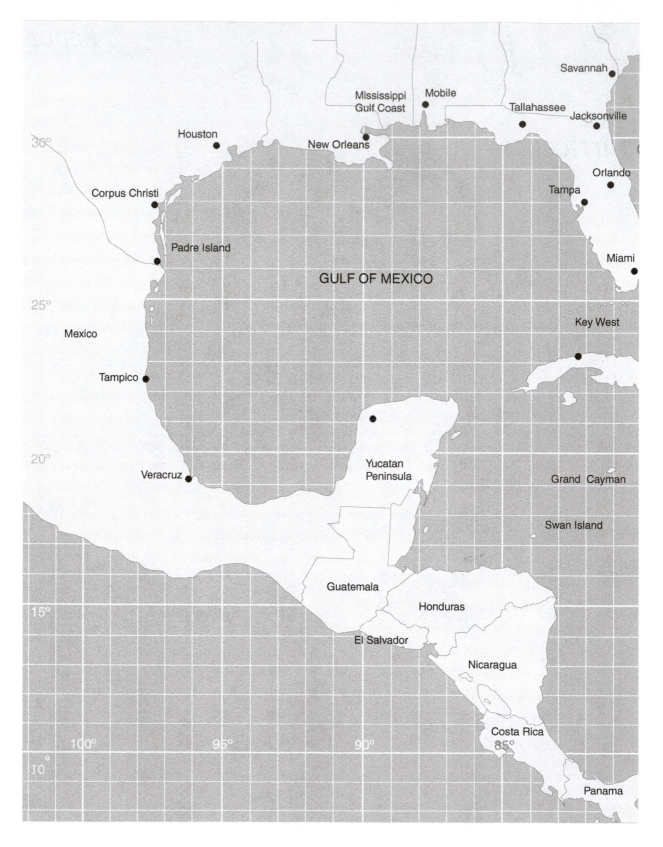

Figure **13-4** A modified NOAA hurricane tracking chart.

3. Using the data in Table 13-5, plot the location of the eye of Hurricane Zeke during July 15 and 16 on Figure 13-4.

TABLE 13-5

Hurricane Zeke—July 1998

Date	Time	Latitude (°N)	Longitude (°W)	Wind speed (miles/hour)	(kilometers/hour)
7/15	0000	26.2	78.6	120	200
	0800	27.9	79.1	115	192
	2000	29.8	80.0	110	183
7/16	0000	30.7	80.3	120	200
	0800	31.6	80.5	110	183
	1015	32.0	81.0	110–Landfall	183

(a) Compare your estimated landfall from Question 2 to the true landfall. How accurate (day and time) was your estimate? _____

(b) Compare your hurricane watch and warning predictions to the ultimate landfall. Did Hurricane Zeke make landfall at your predicted geographic location? _____

(c) If Hurricane Zeke did *not* make landfall at your predicted location, why not? (What atmospheric or oceanographic factors might have affected the storm's path?) _____

_____

_____

_____

4. Table 13-6 is a shoreline survey conducted prior to the landfall of Hurricane Zeke. These data show distance perpendicular to shoreline from a starting point ("benchmark"), and the elevation difference from the benchmark. Plot these data on the graph on the following page.

TABLE 13-6

Shoreline survey of Noname Beach

Distance seaward from benchmark (meters)	(feet)	Elevation relative to mean sea level (meters)	(feet)
0	0	1.5	5.0
4.5	15	1.4	4.6
11.1	35	1.2	4.1
15	50	1.3	4.4
23	75	1.5	4.9
27	90	1.0	3.2
30	100	0.7	2.2
38	125	1.2	0.5
41	135	0.0	0.0
46	150	0.6	−1.9
53	175	1.1	−3.4
61	200	1.1	−3.6

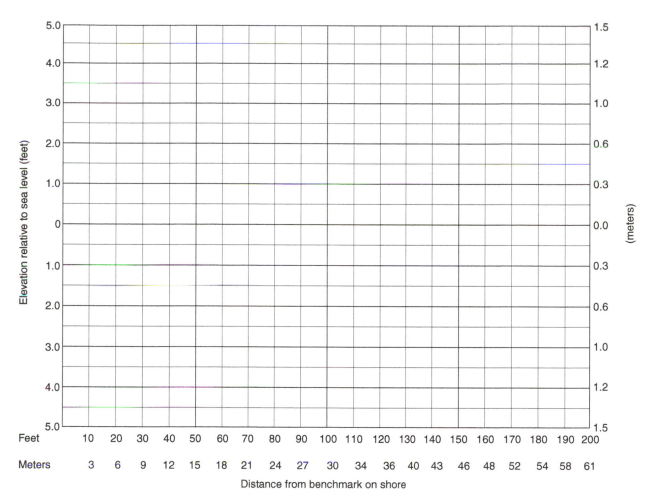

Feet 10 20 30 40 50 60 70 80 90 100 110 120 130 140 150 160 170 180 190 200

Meters 3 6 9 12 15 18 21 24 27 30 34 36 40 43 46 48 52 54 58 61

Distance from benchmark on shore

5. Table 13-7 is a shoreline survey conducted *after* the landfall of Hurricane Zeke. Data are shown relative to the same benchmark previously described.

TABLE 13-7

Post-storm shoreline survey of Noname Beach

Distance seaward from benchmark		Elevation relative to mean sea level	
(meters)	(feet)	(meters)	(feet)
0	0	1.5	4.8
4.5	15	1.5	5.0
7.6	25	0.8	2.5
12	40	0.5	1.5
15	50	0.2	0.7
23	75	0.1	0.3
24	80	0.0	0.0
30	100	0.3	− 1.0
38	125	0.6	− 1.9
46	150	0.4	− 1.2
53	175	0.5	− 1.5
61	200	0.6	− 2.0

(a) Using a different color, plot these "post-hurricane" data on the graph in Question 4.

(b) What effect did the hurricane have on the position of the shoreline and the profile of the beach? ____

_____

_____

Where was sand eroded from, and to where was this eroded sand moved? _____

_____

_____

6. Figure 13-5 is a map of the Mississippi coastline affected by Hurricane Camille in 1969, with place names numbered and survey stations labeled. Table 13-8 lists storm surge data for the inundated coastal area of Mississippi.

(a) Transfer the data from Table 13-8 to Figure 13-5, then draw an east-west profile along the coastline, depicting storm surge height along coastal Mississippi.

TABLE 13-8

Storm surge data from Hurricane Camille, August 17, 1969

Location	Height*	Location	Height*
Pearlington		11	20.4
1	11.2	12	19.9
2	16.2	13	19.5
Waveland	18.5	14	18.5
Bay St. Louis	21.7	Biloxi	
3	17.2	15	16.0
30	12.1	25	13.9
29	13.8	26	12.5
4	19.2	24	3.3
5	17.6	16	14.2
6	22.6	17	15.7
Pass Christian		Ocean Springs	15.8
7	24.2	18	14.0
8	21.0	19	13.3
9	20.2	20	12.7
Gulfport		21	8.6
28	13.7	Pascagoula	11.2
27	14.3	22	9.2
10	20.8	Bayou La Batre	8.5

*Data in feet above mean sea level.

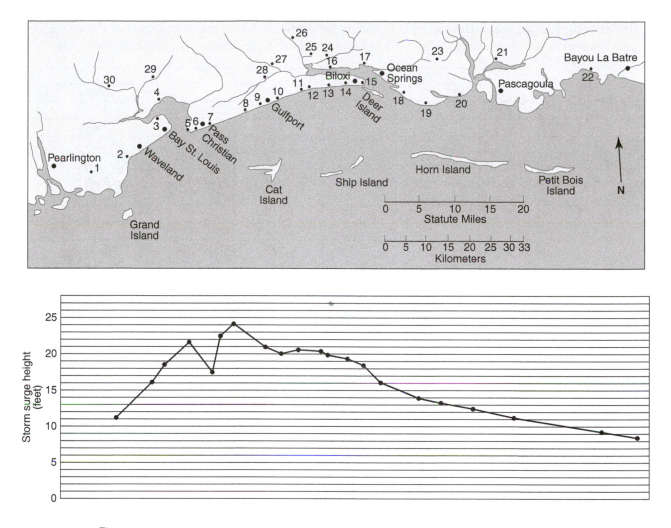

Figure 13-5 A chart of the Mississippi coastline affected by Hurricane Camille in 1969. Place names and numbered measurement locations are shown. Storm surge heights are provided in Table 13-8.

(b) Since the highest wind stresses in a cyclonic storm will be in the northeast quadrant of the storm, does the maximum storm surge height give you information on the landfall of the eye of Hurricane Camille? Explain. _____

_____

_____

_____

_____

(c) Before the storm, Ship Island was one continuous east-west-trending barrier island. Does the post-storm condition of Ship Island provide any information about the storm's track and its landfall? Explain.

_____

_____

_____

_____

_____

Oceanography from Space and Remote Sensing

OBJECTIVES:

■ To understand the various uses of remote sensing systems.

■ To map sea-surface temperatures and the extent of warm and cold water masses.

■ To map the important areas of phytoplankton productivity.

*R*emote sensing may be defined as the collection of information or data about objects without actually being in physical contact with them. In 1858, just 32 years after the invention of photography, the first aerial photograph was taken from a tethered balloon. The first photograph from a rocket was taken in about 1906, and Wilbur Wright took the first photograph from an airplane in 1909. The first artificial satellite to orbit the earth was Sputnik I, launched by the Soviet Union in 1957. Four years later, the United States launched a weather satellite, Tiros 1. Tiros is an acronym for Television and Infrared Observation Satellite. Almost all subsequent earth-orbiting satellites were named following this convention; for instance, the Earth Observation Satellite is known as EOS. In 1965 there were 3000 detectable earth-orbiting objects, sometimes called "space junk," and in 1998 this number had tripled to 9000 objects. Many of these objects are weather, earth-monitoring, or communication devices.

Electromagnetic Radiation

Remote sensing uses instruments or devices that record reflected and radiated electromagnetic energy from the earth's surface. **Electromagnetic radiation** (EMR) has no mass, but hot bodies such as our sun emit huge amounts of radiant energy. Sensible heat and visible light are part of this energy emission. Artificial devices such as microwave ovens, light bulbs, and heat lamps emit the same kind of energy, but on a much smaller scale. EMR is emitted by all bodies (including yours) that are at a temperature above absolute zero ($-273°C$ or $-459°F$). The temperature of a body determines the nature of its emitted radiant energy. Bodies with a high surface temperature emit high-energy radiation with short wavelengths (penetrating radiation, like x-rays), and those with a cooler surface emit lower-energy radiation at longer wavelengths (**infrared** radiation or heat rays). Even an ice cube emits some radiant energy. In other words, EMR is a spectrum or continuum of wave energy that runs from very short, high-energy gamma and x-rays to very long, low-energy radio waves (Figure 14-1). Some fundamental properties of electromagnetic radiation pertinent to our understanding of remote sensing are:

1. It travels at the speed of light (300,000 kilometers/second or 186,000 miles/second)

2. It has wave characteristics analagous to those of ocean waves (see Exercise 11)

3. It can travel in a vacuum, in contrast to sound waves, which need a medium in order to be transmitted.

Remote sensing systems, therefore, employ EMR reading devices such as film cameras, infrared sensors, and microwave and radar sensors.

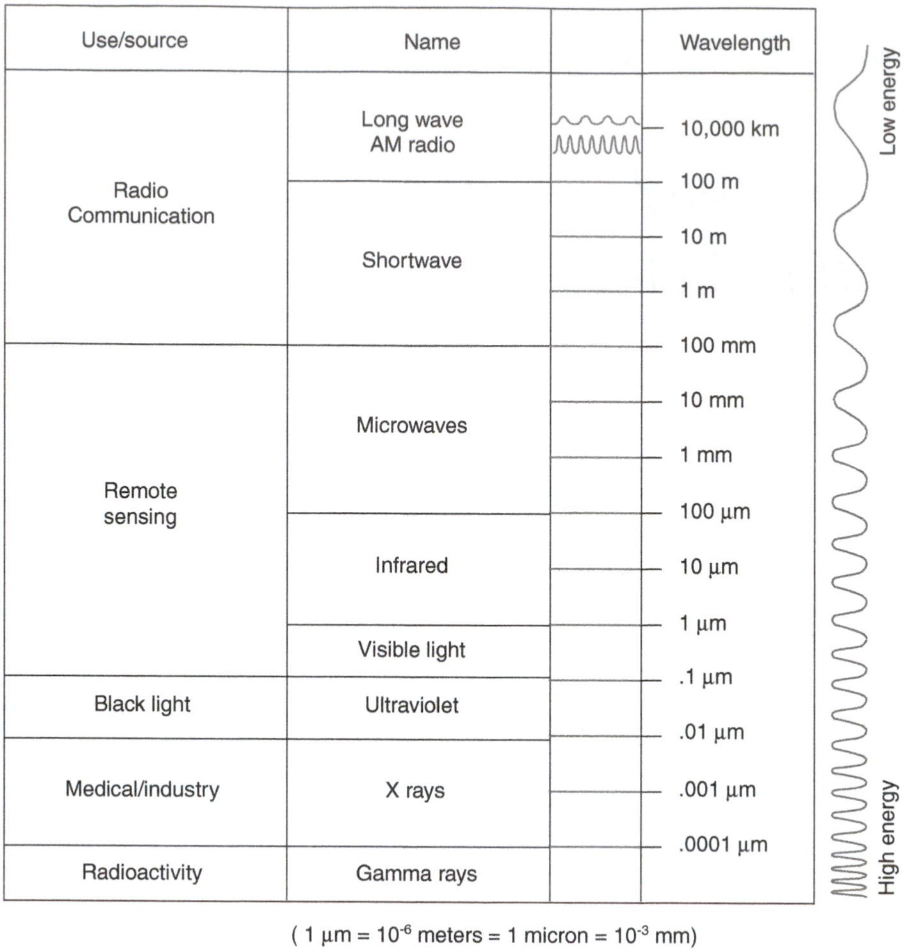

Use/source	Name		Wavelength	
Radio Communication	Long wave AM radio		10,000 km	Low energy
			100 m	
	Shortwave		10 m	
			1 m	
Remote sensing			100 mm	
	Microwaves		10 mm	
			1 mm	
	Infrared		100 μm	
			10 μm	
	Visible light		1 μm	
Black light	Ultraviolet		.1 μm	
			.01 μm	
Medical/industry	X rays		.001 μm	
			.0001 μm	
Radioactivity	Gamma rays			High energy

(1 μm = 10^{-6} meters = 1 micron = 10^{-3} mm)

Figure 14-1 The electromagnetic spectrum, and some scientific and peaceful uses of this form of wave energy.

Sensing the Earth

The sun bathes the earth with light and invisible forms of radiant energy. Some of this energy fuel the biosphere, some is reflected, and some is absorbed by rocks, soil, atmosphere, and oceans (Figure 14-2). Much of this absorbed energy is then radiated back into space in the infrared (IR) part of the spectrum. Figure 14-3 illustrates semi-quantitatively how the hot sun and the cool earth radiate energy at wavelengths important to remote sensing. Note that all of the earth's absorbed solar energy is emitted in the infrared region. Also note that the amount of solar energy reaching the earth from the sun must equal the amount emitted by the earth, or global warming or cooling will result. When less heat is lost to space than is received

from the sun, then global warming is the consequence. The slightly higher solar energy input (sun at earth curve) in Figure 14-3 reflects photosynthesis and storage of solar energy by plants. Long-term storage is in the form of coal and oil, which have a plant origin.

It is estimated that the oceans lose 40 percent of the solar energy they receive. Because they back-radiate in the long-wavelength portion of the spectrum, sensors that detect IR radiation are very useful in obtaining remote images of the oceans (Table 14-1).

An important measure for the quality of visible-light remote images of earth is **albedo,** the percentage of visible radiation that is reflected away from an opaque (solid earth) or translucent (atmosphere and ocean) surface. Seen from

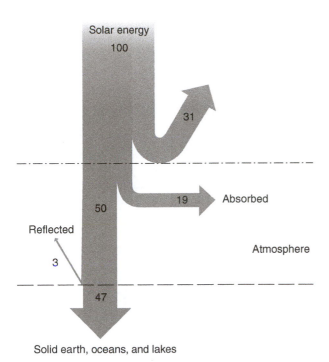

Figure 14-2 The amount of solar energy reaching the surface of the earth on a day with average cloud cover. About half the solar energy that encounters the earth's upper atmosphere reaches the surface of the earth; the rest is reflected or absorbed by the atmosphere. [After Robert H. Romer, *Energy; An Introduction to Physics.* W. H. Freeman and Company. Copyright © 1976.]

TABLE 14-1	
The spectral distribution of the sun's energy and the earth's backradiation energy	
Type of radiation	Percentage of total energy
Sun	
Infrared and heat rays	50.0
Visible light rays	40.5
X-rays and gamma rays	9.0
Ultraviolet rays	0.5
Earth	
Infrared and heat rays	100.0

niques that depend upon visible light use film or video cameras.

Most satellites (spy satellites excluded) are meteorological ones. These satellites tell us a great deal about the movement and temperature of the atmosphere and oceans, and aid greatly in weather forecasting. The information gleaned is invaluable in planning, particularly for air and sea transportation. A satellite's sensors are called **radiometers,** and they are designed to measure radiation brightness in very narrow wavelength ranges known as **channels.** An important sensor for measuring sea-surface temperature (SST) is the Advanced Very High Resolution Radiometer (AVHRR, Color Plates 1 and 3). It measures IR radiation from the earth at five wavelengths (channels) that are able to

space, bright objects, such as clouds and snow, have high albedos, and low-reflectivity objects, such as the oceans, some forests, and lakes, have low albedos (Table 14-2). Remote-sensing tech-

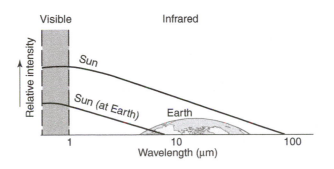

Figure 14-3 The sun's and earth's relative emissions of electromagnetic radiation at wavelengths of interest for remote sensing. Note that all the earth's emissions are in the infrared region of the spectrum.

TABLE 14-2

Albedos expressed as the percentage of
visible light reflected from surfaces for
selected earth features in satellite imagery

Feature	Albedo
Clouds	30–92
Fresh snow	75–90
Week-old snow	40–70
Sand dunes, dry	35–45
Concrete	17–20
Blacktop roadway	5–10
Oceans and lakes	7–9
Sunglint on Gulf of Mexico	17
Desert	25–35
Tundra	15–20
forest, deciduous	10–20
Forest, coniferous	5–15

Adapted from Eric D. Conway and the Maryland Space Grant
Consortium, *An Introduction to Satellite Image Interpretation.* Johns
Hopkins University Press, Baltimore, 1997.

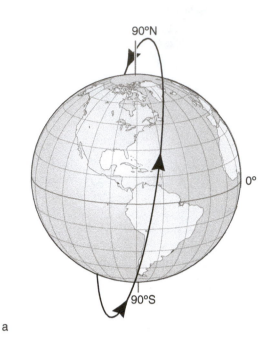

a

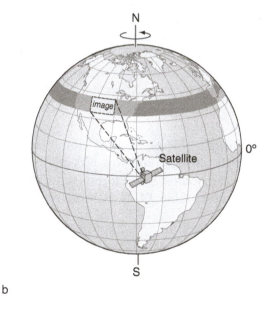

b

Figure 14-4 Orbits for earth-imaging satellites. (a)
The polar orbit. The orbital plane is tilted 8° to the poles
and is fixed in space as the earth rotates under the satellite. With several satellites in polar orbit, the entire earth
may be seen and remotely imaged. (b) A geostationary
orbit allows a satellite to remain over the same point on
the earth throughout its entire orbit. Because the earth
does not move in relation to the satellite, a time series of
images of a particular area on the earth can be obtained.

penetrate the atmosphere. The visible (VIS) channel on the AVHRR measures the amount of solar
radiation that is reflected away from the earth.
Since earth features tend to have different albedos,
interpreters with experience are able to distinguish
between cloud types, water, ice, rocks, and soil.

For over 100 years marine biologists have studied and tried to understand the factors controlling
the distribution, growth, and fate of phytoplankton. Because these microscopic floating plants are
the base of the marine food chain, their growth
and distribution (productivity) are important in a
growing world partially dependent upon the sea
for food. Adequate phytoplankton sampling of
the vast ocean is not possible using ships alone.
The Coastal Zone Color Scanner (CZCS), first
launched on Nimbus-7 in 1978, takes advantage of
the fact that phytoplankton contain chlorophyll
and other pigments that capture sunlight (see
Color Plates 4 and 5). As phytoplankton increase in
number, their associated chlorophyll pigments absorb blue light, and water color changes from blue
to green. The change in the ratio of blue to green
enables us to compute chlorophyll content. First-
generation CZCS covered an ocean swath 1600
kilometers (1000 miles) wide with a **pixel** resolution of 0.8 kilometers (0.5 miles).

Importance of the Orbit

The National Oceanic and Atmospheric Administration (NOAA) operates most of the United States satellites that are useful in studying the atmosphere and oceans. Satellites designed to study weather, oceans, or the solid earth have one of two orbits around the earth: *polar* or *geostationary.* Polar orbits are from pole to pole with an inclination of 98°; that is, the satellite passes within 8° of the poles on every orbit (Fig. 14-4a). These satellites are *sun-synchronous;* that is, the orbital plane of the satellite remains stationary with respect to the sun. As the satellite follows its orbit, the earth rotates below it. Each orbit has a period of about 102 minutes, which means that a satellite makes roughly 14 orbits per day. As you can surmise from this description, these satellites have excellent coverage of the entire earth's surface. The United States maintains at least two polar-orbiting satellites at any one time.

Geostationary satellites orbit the earth at an altitude of about 35,000 kilometers or 21,000 miles above the equator. At this altitude, the **angular velocity** of the earth and the angular velocity of the spacecraft are equal. Thus, each makes a complete revolution of 360° in 24 hours, and the satellite stays over the same point on the earth all the time (Figure 14-4b). Thus a time-lapse sequence of images of the same area on the earth can be assembled (see Color Plate 2). This type of satellite is useful for monitoring forest fires in Brazil and Russia, for instance. Geostationary Operational Environmental Satellites (GOES) are located over the equator at 75°W and 135°W longitude, and provide overlapping coverage of the entire North and South American continents as well as the Atlantic and Pacific Oceans.

DEFINITIONS

Albedo. A property that describes how much light a surface, object, or area reflects

Angular velocity. The rate at which a body moves along an arc of a circle, expressed in radians/second or revolutions/minute.

Channel. A specific band of wavelengths sensed by a satellite sensor.

Electromagnetic radiation (EMR). Energy in wave form emitted by all objects above absolute zero.

Infrared (IR) radiation. That portion of the electromagnetic spectrum adjacent to the long-wavelength or red end of the visible light range. Invisible to the eye, it can be detected as a sensation of warmth on the skin (sensible heat). Most energy released by moderately heated surfaces is infrared.

Pixel. The smallest element in a satellite image. Thousands of pixels make up each image and when viewed together make a picture.

Radiometer. An instrument that measures the intensity of electromagnetic radiation at specific wavelengths. Satellite radiometers measure infrared radiation and visible light.

Exercise 14

Oceanography from Space and Remote Sensing

NAME _____

DATE _____

INSTRUCTOR _____

1. The sea-surface temperature (SST) image in Color Plate 3 represents the average SST taken over a 12-year period.

   (a) Explain the wedge-shaped mass of cool water along the equator off the west coast of South America. The tiny dots in the mass are the Galápagos Islands, which lie at 0° latitude, and the water there is quite cool and pleasant. There are two rather straightforward reasons for the colder water, which are:

   i. _____

   _____

   ii. _____

   _____

   (b) What is the average SST off San Diego, California, on the west coast of North America (_____°C), and Savannah, Georgia, on the east coast (_____°C), which are at approximately the same latitude north of the equator? Explain this difference. _____

   _____

   _____

   _____

   _____

2. Color Plate 4 shows upwelling on the west coast of the United States (California and Oregon).

   (a) From which direction(s) must the wind blow to produce the upwelling and cool water seen in Color Plate 4b. Circle the best answer below, keeping in mind Ekman transport and currents.

   North to south?   East to west?   South to north?   West to east?

   (b) Why doesn't the cold upwelled water parallel the coast all the way from Oregon to the southern limit of the photo?

   _____

   _____

   _____

   _____

   _____

   (c) Explain the plume of warm water off San Francisco Bay.

   _____

   _____

   _____

   _____

   _____

(d) In Color Plate 4a, note how the distribution of high chlorophyll (phytoplankton) concentrations at (8) matches the areas of intense upwelling in Color Plate 4b.
Briefly explain why this match is predictable. _____
_____
_____
_____
_____

3. Color Plate 5 shows chlorophyll concentrations in the western North Atlantic Ocean.
(a) An interesting feature of the Gulf Stream is the formation of rings (1, 2, 3), which pinch off from meanders in the main flow. A recently formed warm-core ring with Sargasso Sea water in the center can be seen east of Delaware Bay (1). What can be said of the fishing possibilities in the Sargasso Sea?
_____
_____
_____
_____
_____

(b) What can be said of the commercial fishing potential of Nantucket Shoal (4) and Georges Bank (5)?
_____
_____
_____
_____

(c) The chlorophyll-rich banks at (6) are known as the _____. How may one explain the high productivity of these shallow banks that rise from very deep water?_____
_____
_____
_____

(d) Does the image suggest that the Great Lakes are highly productive?_____
_____
_____
_____
_____

Distribution of Marine Life

OBJECTIVES:

■ To learn about the different life zones of the ocean.

■ To appreciate the physical and chemical factors that control the distribution of life within and between these zones.

In this exercise we investigate how and why the distribution and abundance of organisms varies within the ocean. While life in some form may be found in every part of the marine realm, even in the deepest trenches and most oxygen-depleted waters, the physical and chemical conditions of different environments play a fundamental role in fostering or excluding particular forms of life. The natural breaks between these different environments may be used to define marine life zones, which provide a useful framework to compare and contrast marine diversity as a whole.

Marine Life Zones

The most commonly used classification of marine life zones was defined by Joel Hedgepeth and other biologists and is based on natural breaks between major assemblages of marine organisms, or **biomes** (Figure 15-1). The marine environment is first divided into two fundamentally different zones: pelagic and benthic. The **pelagic zone** consists of the entire water column and is further subdivided into the *neritic zone* (water overlying continental shelves) and the *oceanic zone* (water off of continental shelves). Given the great depth and environ-

mental range of the oceanic pelagic zone, it is further subdivided by depth into the *epipelagic zone*, which generally coincides with the *photic zone*, or depth to which light effectively penetrates into the ocean, and the progressively deeper *mesopelagic*, *bathypelagic*, and *abyssopelagic zones*.

The **benthic zone** consists of the entire sea floor. It is finely subdivided from above the shoreline seaward into (1) the *supralittoral zone*, kept moist by sea spray; (2) the *littoral zone*, between the highest and lowest tides; (3) the *inner sublittoral zone*, from lowest low tide to the depth at which plants can no longer photosynthesize; (4) the *outer sublittoral zone*, from the base of the inner sublittoral zone to about the continental shelf edge or slope break; (5) the *bathyal zone*, from the shelf break to about 4000 meters depth; (6) the *abyssal zone*, from the base of the bathyal zone to the deepest depths of the abyssal plains; and (7) the *hadal zone* of deep waters found within oceanic trenches.

The divisions of these zones are based on observed natural breaks in the distribution of marine organisms. For example, marine biologists have documented significant and consistent changes in marine life occurring at depths of 100–200 meters. Thus, this depth was defined as the break between the epipelagic and mesopelagic zones. As with most attempts to impose a rigid framework on the natural world, the Hedgepeth zonation is not perfect; for example, the epipelagic–mesopelagic boundary is well defined in tropical and temperate seas, but is less pronounced in polar regions, where chemical and physical conditions change less dramatically with depth.

The various organisms that constitute these major biomes can be classified on the basis of how

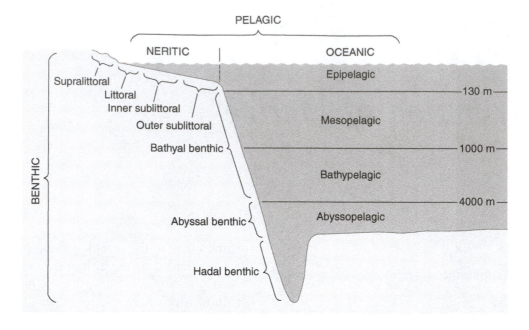

Figure 15-1 Modified Hedgpeth classification of life zones in the ocean. The pelagic environment is divided into neritic and oceanic zones, with the oceanic zone further subdivided into finer zones (approximate depths are given in meters). The benthic environment is also subdivided into finer zones, from supralittoral to hadal benthic.

they interact with their environment. The different descriptive terms for organisms living in the pelagic and benthic zones are as follows. Note that for the benthic zone, more than one term may apply; for example, a worm that burrows through sediment may be termed a vagrant infaunal animal.

PELAGIC ZONE:

Nektonic: Actively swimming organisms (e.g., fish, seals)

Planktonic: Organisms unable to swim effectively against a 1-knot current (e.g., jellyfish)

Holoplanktonic: Organisms that are planktonic throughout their life cycle

Meroplanktonic: Organisms that float during the early part of their life cycle and are either benthic or nektonic for the remainder

BENTHIC ZONE:

Sessile: Attached or immobile organisms (e.g., corals)

Vagrant: Mobile or roving organisms (e.g., crabs)

Infaunal: Organisms living within the sea floor (e.g., most worms)

Epifaunal: Organisms living on or attached to the sea floor (e.g., crabs, corals)

Factors that Control Marine Life Zones

Although the Hedgepeth classification is based on the observation and documentation of major marine biomes, the underlying controls upon these biomes and their boundaries are chemical and physical in nature. To illustrate this concept, we shall describe a hypothetical pattern of distribution for a copepod species (a holoplanktonic crustacean) in terms of the biological, physical, and chemical factors that limit its distribution. This three-dimensional distribution is illustrated in Figure 15-2. The copepod's occurrence is limited by specific temperature, salinity, and biological boundaries, numbered 1–8 in the figure and in the following paragraphs.

Boundaries 1 and 2 are thermal. Our hypothetical copepod is *stenothermal,* or restricted to a narrow range of temperatures (*eurythermal* organisms, in contrast, are tolerant of a relatively wide range of temperatures). A surface current flows through the area, as indicated by the arrow.

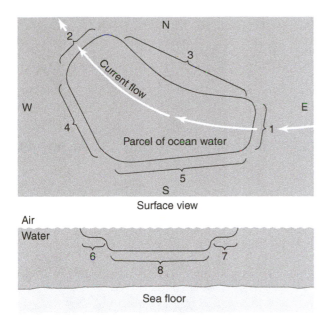

Figure 15-2 Diagram of the distribution of a hypothetical copepod species, with the various boundaries controlling its distribution.

Within boundary 1, seawater temperatures are mild enough to support our copepod (outside this boundary, the temperature is too warm), whereas beyond boundary 2 the water has cooled to a temperature that excludes the copepod.

Boundaries 3 and 4 are biological. Our hypothetical copepod feeds exclusively on diatoms, a type of single-celled phytoplankton, and, in turn, the copepod is fed upon by a predator. Boundary 3 represents the northern limit where diatom abundance becomes too low to support a copepod population. Boundary 4 represents the easternmost habitat boundary of a predator that feeds very efficiently upon copepods, and thus maintains their population at effectively zero west of the boundary.

Boundary 5 is chemical and is marked by a salinity that is too high for the copepod. Our copepod is *stenohaline*, which means it can tolerate only a narrow salinity range (*euryhaline* organisms, in contrast, are tolerant of a wide range of salinities).

Boundaries 6 and 7, like boundaries 1 and 2, are defined by temperature. Below these boundaries, temperatures are too cool for the copepod, although its diatom food source is plentiful in these cooler waters.

Boundary 8 is biological, and relates to the deepest depths at which diatoms receive sufficient light to photosynthesize. Thus, while copepods could theoretically exist below this boundary, there is no food to support them.

This environmental framework defines the geographic range of our hypothetical copepod. Note that changes in the environment, such as those occuring over a seasonal cycle, would also alter the geographic range of the copepod. Oftentimes, the same type of organism may be found in two or more separate geographic regions; for example, some organisms are restricted to the north and south polar regions. This pattern is known as a *disjunct* distribution.

DEFINITIONS

Benthic zone. The bottom of the ocean. An organism that lives in the benthic zone lives on or in the ocean bottom.

Biomes. Large natural assemblages of organisms. On land, for example, a biome could consist of all organisms living within a tropical rain forest.

Pelagic zone. The ocean water column from the surface to the bottom. An organism that lives in the pelagic environment lives in the water column.

Exercise *15*

Distribution of Marine Life

NAME _____

DATE _____

INSTRUCTOR _____

1. Answer the following questions by stating the physical or chemical factors that could potentially explain the various divisions of the Hedgepeth classification. Be specific and scientific in your responses; for example, "the bottom of the seasonal thermocline" (see Exercise 7) or "the bottom of the photic zone" (see Exercise 16) could be appropriate answers. Consult other laboratory exercises, lecture notes, and your textbook to review the physical and chemical variations within the ocean. Depending upon what aspects of oceanography you have already covered in your course, your instructor will tell you how many factors to identify for each question.

(a) Division between epipelagic and mesopelagic zones. _____

_____

_____

_____

(b) Division between mesopelagic and bathypelagic zones. _____

_____

_____

_____

(c) Division between bathypelagic and abyssopelagic zones. _____

_____

_____

_____

(d) Division between abyssal benthic and hadal benthic zones. The bottom waters bathing both of these environments are essentially the same; therefore, consider why trench benthic faunas might differ from those living on the abyssal plains. (Hint: Examine the distribution of trenches and think in terms of biological evolution over the course of geologic time.) _____

_____

_____

_____

(e) Division between supralittoral and littoral zones (see also Exercise 22). _____

_____

_____

_____

(f) Division between outer sublittoral and bathyal zones. Cite factors different from those you have already given for the division between the epipelagic and mesopelagic zones in (a). _____

_____

_____

_____

(g) Division between bathyal benthic and abyssal benthic zones. (Hint: See Exercise 5 to review the sediment changes that occur at approximately this depth.) _____

_____

_____

(h) Why is there no 1000-meter boundary in the benthic environment, but a significant one at about the same depth in the pelagic environment? (Hint: What is a major control upon benthic organisms that pelagic organisms would not be affected by, and vice versa?) _____

_____

_____

2. Some marine species exhibit a disjunct geographic distribution termed *bipolar,* living in shallow waters around the North and South poles, but not in shallow or deeper waters between the poles. State a few possible biological or physical reasons to explain this type of distribution. _____

_____

_____

_____

3. Another kind of distribution is *tropical submergence,* in which a species lives in shallow polar waters, but at depth in temperate and tropical regions. State a few biological, physical, and/or chemical reasons that would promote this type of distribution. _____

_____

_____

_____

_____

4. During the 1983–84 El Niño (see Exercise 17), the pronounced warming of equatorial Pacific waters led to the death of many corals that lived within the shallow photic zone. Oftentimes, deeper-dwelling coral species suffered greater mortality from this warming than did shallow-water species. Propose some hypotheses to explain this pattern of coral mortality. _____

_____

_____

_____

_____

What types of data would be required to test your hypotheses? _____

_____

_____

_____

_____

5. The evolution of new species seems to be favored when a subpopulation of the "parent" species becomes geographically isolated and subjected to different environmental conditions over geologic time. Given this generalization, describe the distribution characteristics of a hypothetical "parent" species that would be most favored to produce many "daughter" species over time. _____

_____

_____

_____

Primary and Secondary Productivity

OBJECTIVES:

■ To understand the process of primary productivity and its importance.

■ To examine the physical, chemical, and biological controls upon marine primary productivity, both over the ocean surface and at different depths.

■ To appreciate how secondary productivity relates to primary productivity.

Primary productivity is the conversion of inorganic materials into organic compounds. Organisms that perform this amazing feat are termed **autotrophs** ("self-feeding"). The most important form of primary productivity is **photosynthesis,** in which solar energy is used to convert carbon dioxide and water into simple carbohydrates and other organic molecules. Essentially all of the free oxygen in the atmosphere is derived from photosynthesis, and at least half of all photosynthesis takes place in the oceans. The carbohydrate produced by photosynthesis forms the base of the marine food chain.

The Process of Photosynthesis

The net chemical reaction for photosynthesis is

$$6\,CO_2 + 6\,H_2O \longrightarrow C_6H_{12}O_6 + 6\,O_2$$

Carbon Dioxide   Water   Sunlight   Carbohydrate   Oxygen

In reality, this chemical process is quite complex. It can be divided into two distinct components, light reactions and dark reactions. In light reactions, so-lar energy is intercepted by and excites photosynthetic pigments (mainly chlorophyll) within cells, which pass this energy on in a form available for dark reactions. During dark reactions, the energy from light reactions is used to convert carbon dioxide and water molecules into free oxygen and "fixed" carbon in the form of carbohydrate. Dark reactions are so called because they can occur in darkness, unlike light reactions.

The carbohydrate created by photosynthesis is used as a building block for more complex organic compounds or as an energy source. The latter is accomplished by reversing the process of photosynthesis, whereby the energy contained within the carbohydrate's high-energy bonds is used to produce ATP, a molecule used to "power" the organism. This reverse process uses free oxygen to "burn" the carbohydrate and is termed **respiration;** all organisms respire.

$$C_6H_{12}O_6 + 6\,O_2 \longrightarrow 6\,CO_2 + 6\,H_2O$$
+ sufficient energy to synthesize
36 ATP molecules

Another type of primary productivity is *chemosynthesis,* in which bacteria break down iron- and sulfur-containing compounds and use the energy derived from their chemical bonds to convert carbon dioxide and water into carbohydrates and oxygen. Such nonphotosynthetic reactions probably account for only a small percentage of global primary productivity, but can be locally important in environments where photosynthesis is not possible, such as the deep ocean. A primary example of chemosynthesis occurs around hydrothermal vents at oceanic spreading centers, where bacteria oxidize the hydrogen sulfide of vent fluids into sulfur and sulfate to fix carbon. This process supports a rich and unique ecosystem at depths of 2000 meters and greater.

While autotrophs have the ability to produce their own food, **heterotrophs** ("other-feeding") cannot, and must consume autrophs or other heterotrophs to obtain their energy. This transfer of autotroph-produced organic compounds from one **trophic level** to another is known as **secondary productivity**.

Quantifying Marine Productivity

Productivity, be it primary or secondary, is generally measured as the rate of carbon fixation and expressed in units of grams of fixed carbon (dry weight) per square meter of ocean surface per year (g $C/m^2/yr$). There are a number of methods for measuring this process; the two most common are described below. Note that these and other methods all use the basic photosynthesis equation as a guide to determining the amount of productivity.

1. Light–Dark Bottle Method. Seawater is collected from a given depth along with its natural component of plankton, although sometimes the seawater is screened to exclude larger zooplankton. The dissolved oxygen concentration is measured, and the water sample is split into two bottles, one transparent ("light bottle") and one opaque ("dark bottle"). The two bottles are then returned to the original sampling depth. After a specific interval of time (e.g., 2 hours), the bottles are retrieved, and the dissolved oxygen concentration in each sample is re-measured. These new concentrations are then subtracted from the initial concentration to provide a measure of the change in dissolved oxygen concentration in each bottle over the time interval. Thus, the rate of photosynthesis can be determined in two forms: **gross photosynthesis** (total amount of oxygen produced through photosynthesis) and **net photosynthesis** (amount of oxygen produced through photosynthesis minus the amount of oxygen consumed through respiration).

2. "Hot Carbon" or ^{14}C Method. A known amount of radioactive or "hot" carbon (^{14}C) in the form of bicarbonate is added to a seawater sample in a transparent bottle. The sample is returned to the original depth for a specific interval of time (again, generally 2 hours), during which photosyn-

thesis continues. The sample is then retrieved and filtered to collect all the plankton. Productivity is then determined by measuring the amount of hot carbon incorporated into the plankton compared with its concentration in the bottle over the time interval.

Distribution of Primary Productivity

The first-order control upon primary productivity in the open ocean is the abundance of phytoplankton living within the **photic zone**—that part of the water column in which phytoplankton receive enough light for photosynthesis. Phytoplankton abundance may be expressed as either the number (standing crop) or biomass (living mass) of phytoplankton living within a square meter area extending from the surface to the base of the photic zone. Phytoplankton abundance is affected by (1) the *availability of light,* which varies with degree of latitude, time of year, time of day, clarity of the water, and the number of organisms in the water ("biological overshadowing"); (2) the *availability of limiting nutrients,* such as phosphates and nitrates (see Exercise 18); and (3) the *rate of grazing* by herbivores, such as copepods and other zooplankton. Because of these many factors, productivity varies geographically over the ocean's photic zone as well as vertically within any given spot. Color Plate 8 shows the global distribution of primary productivity in surface waters through satellite imagery of chlorophyll concentrations, the predominant photosynthetic pigment.

Of these three factors, the principal ones are the availability of light and nutrients. Whereas light availability near the ocean's surface is a limiting factor only at high latitudes, light does become a limiting factor with increasing water depth. This decrease in light intensity occurs through both light reflection, due to plankton and suspended sediments, and light absorption by ocean water. The **photic zone** extends from the surface to the depth at which photosynthesis is less than 1 percent of that at the surface. In the open ocean, the photic zone generally extends to a depth of 100–200 meters (Figure 16-1).

In general, nutrient availability is a much more significant factor than light in controlling productivity within the photic zone. All organisms require certain nutrients, such as nitrogen (N) and phos-

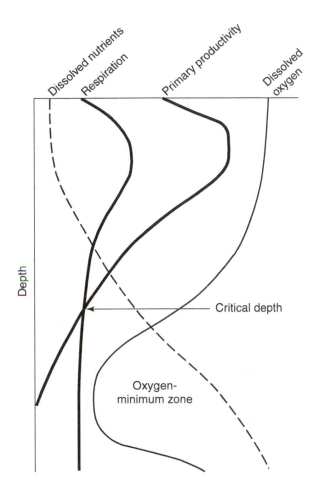

Figure 16-1 Generalized diagram of relative changes in primary productivity, respiration, dissolved oxygen, and nutrients. Note that the critical depth occurs where primary productivity equals respiration, and that primary productivity is effectively zero below the base of the photic zone. The oxygen-minimum zone is produced by respiration of organic matter as it sinks from the photic zone.

phorus (P), in addition to the carbohydrates produced by photosynthesis. Autotrophs extract these elements from seawater, and heterotrophs acquire them by consuming autotrophs and other heterotrophs. These nutrients are supplied to the ocean through chemical weathering of continental rocks. Biological nutrient recycling within the ocean, discussed below, maintains a much higher ocean nutrient content than would otherwise be expected. Within the photic zone, however, most nutrients are depleted (i.e., limited or non-conservative) because autotrophs rapidly incorporate them into their bodies as they grow. Furthermore, when organisms die, most of the nutrients they contained are rapidly recycled by living organisms.

This nutrient recycling produces a vertical nutrient concentration pattern, as shown in Figure 16-1. Below the photic zone, nutrient concentrations increase because some organic material (e.g., waste, dead organisms, shell molts) sinks below the photic zone before it can be recycled. As some primary productivity occurs over the entire ocean surface and some organic material is always sinking below the photic zone, nutrients build up to relatively high concentrations in bottom waters, but tend to be limited in most surface waters.

The photic zone is replenished with nutrients by three mechanisms: (1) river and atmospheric input of weathered elements from the continents, (2) convective mixing of deep waters and surface waters, and (3) upwelling of nutrient-rich deep waters. Of these mechanisms, upwelling is the most important source of nutrient replenishment (>90 percent) through its return of nutrients previously exported from the photic zone to deep waters, where they are transported by thermohaline circulation (see Exercise 8). Upwelling is discussed at length in Exercise 7 with the major processes illustrated in Figure 7-4. Thus, the distribution and intensity of primary productivity is strongly controlled by the distribution and intensity of upwelling.

Vertical Variations in Production, Respiration, and Dissolved Oxygen

The rate of primary production within an individual phytoplankton will vary with its depth. Intense sunlight in the shallowest portion of the photic zone may actually inhibit photosynthesis, and progressive darkening with depth decreases photosynthesis. Photosynthesis within an individual phytoplankton is therefore greatest at some intermediate depth, while its respiration rate is relatively constant and depth-independent. At a depth termed the **compensation depth,** an individual phytoplankton's *instantaneous productivity* will be equal to its *instantaneous respiration* (i.e., all of its productivity will be used to meet its most basic energy needs, so that it cannot grow or reproduce). Thus, a phytoplankton must remain above its compensation depth for the majority of its life.

Considering the entire biomass of the upper ocean surface, photosynthesis occurs mainly in the upper photic zone, while respiration by autotrophs, heterotrophs, and decomposers occurs throughout the water column (Figure 16-1). The depth at which total productivity and total respiration are equal is termed the **critical depth.** Above the critical depth, dissolved oxygen concentrations are relatively high because total productivity exceeds respiration. With increasing depth, photosynthesis rates decrease, but heterotroph and decomposer respiration rates remain high. This shift toward a respiration-dominated environment consumes the increasingly limited dissolved oxygen and results in an **oxygen-minimum zone** (Figure 16-1). This zone is generally below the photic zone, but its exact position depends upon water clarity and the vertical variations in the magnitudes of primary productivity and respiration.

Eutrophication

Is increasing the nutrient content in an ecosystem always better because it increases total productivity? The answer depends on what is meant by "better." Increasing nutrient concentrations will generally increase productivity, but this increase may disrupt the natural balance of the existing ecosystem, causing key species to disappear or the entire ecosystem to collapse. Regions of high productivity are termed **eutrophic,** whereas regions of low productivity are termed **oligotrophic.** Variations in nutrient input through upwelling, terrestrial runoff, and vertical mixing strongly influence whether a region is eutrophic or oligotrophic. Human activity can shift coastal waters from a natural oligotrophic state to an altered eutrophic state by dumping raw sewage, agricultural and industrial waste, and so forth into coastal waters (one could think of this as "inwelling" from land). Such *eutrophication* can lead to native organisms being outcompeted by invaders that are better adapted to high-nutrient environments. In some cases, this may lead to the collapse of the entire native ecosystem. An example is the Florida Keys, where increased human nutrient input, mainly in the form of sewage, has greatly increased coastal nutrient levels and led to the overgrowth of oligotrophic-adapted corals by eutrophic-adapted algae.

DEFINITIONS

Autotrophs. Organisms that can produce their own food from inorganic materials.

Compensation depth. Depth at which an individual phytoplankton's instantaneous productivity will be equal to its instantaneous respiration.

Critical depth. Depth at which total productivity and total respiration are equal.

Eutrophic. Having relatively high primary productivity, as in coastal areas with high upwelling or runoff rates.

Gross photosynthesis. Total amount of photosynthesis by a given organism or within a given area. This value does not include the concomitant respiration by that organism or within that area.

Heterotrophs. Organisms that must acquire their food from other organisms.

Net photosynthesis. Amount of photosynthesis by a given organism or within a given area minus concomitant respiration.

Oligotrophic. Having relatively low primary productivity, as in the middle of ocean surface gyres.

Oxygen-minimum zone. Depth zone in which low productivity and high respiration rates lead to low dissolved oxygen concentrations, generally near the base of or below the photic zone.

Photic zone. Depth to which sufficient sunlight penetrates to allow photosynthesis.

Photosynthesis. The process of converting carbon dioxide and water into carbohydrates and oxygen using energy in the form of sunlight.

Primary productivity. The process of converting inorganic compounds into organic compounds. Photosynthesis is the predominant form of primary productivity.

Respiration. The process of converting organic compounds into usable energy by oxidative burning. Essentially the reverse of photosynthesis.

Secondary productivity. The process of converting external organic compounds into internal organic compounds within an organism. Primary productivity is carried out at the first trophic level by autotrophs, whereas secondary productivity is carried out at all higher trophic levels by heterotrophs.

Trophic level. The general position of an organism within an ecosystem with respect to who eats whom.

Exercise 16

Primary and Secondary Productivity

1. The figure below shows a series of light and dark, which represent bottles at 10-meter depth intervals after 2 hours of submergence. The value in each bottle is the *change* in dissolved oxygen concentration from the initial concentration. Use these values as a proxy for primary productivity and respiration during the time interval.

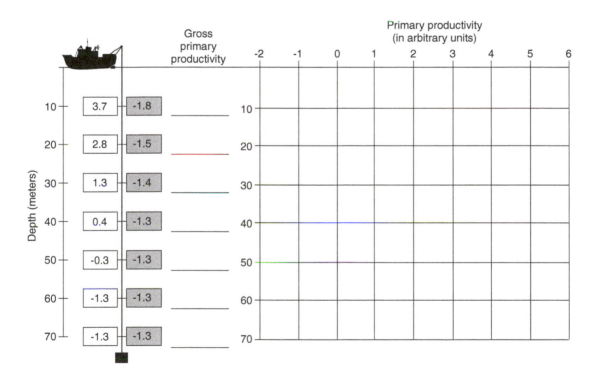

(a) Compared to the original dissolved oxygen concentrations, which bottles (light or dark) tend to have about the same or greater concentrations of dissolved oxygen at shallower depths? _____

Why? _____

_____

_____

_____

(b) Compared to the original dissolved oxygen concentrations, which bottles (light or dark) tend to have lower concentrations of dissolved oxygen? _____

_____

Why? _____

_____

(c) The light bottles reflect the net primary productivity that has occurred at the various depths during the 2-hour interval. Using your responses in (a) and (b) above, write a formula (using LB for light bottle and DB for dark bottle) to determine the gross primary productivity.

Using this formula, calculate the gross primary productivity for each sample in the space provided in the figure. (Hint: Consider what processes are occurring in each bottle.)

(d) On the graph provided, plot the values you calculated in (c) for gross primary productivity and connect them with a solid line, then plot the values for net primary productivity (light bottle values) and connect them with a dashed line. In the surface sample (not shown), the light bottle had a value of 2.4 and the dark bottle had a value of -1.9. Plot these as well. What biological process is represented by the area between the net and gross productivity lines? _____

_____

(e) Draw and label horizontal lines across the graph at the critical depth and at the base of the photic zone. Explain below what each depth represents. _____

_____

What might cause the depth of each to vary over the ocean? _____

_____

(f) In this example, larger zooplankton were not filtered from the seawater samples before resubmergence. How would the removal of larger zooplankton have affected the gross and net primary productivity calculations (i.e., would they be higher/lower)? _____

_____

Why? _____

_____

2. How can you explain the fact that positive gross productivity can take place at greater depths than positive net productivity? _____

_____

_____

3. An *individual* phytoplankton's compensation depth often increases over its lifetime. Why might this occur? _____

_____

_____

_____

4. Environmental conditions strongly control light penetration and therefore critical depths. For the following pairs of environments, indicate which would have the deeper critical depth, and explain your reasoning.

(a) Tropical ocean region far from land vs. polar region far from land _____

_____

_____

(b) Nearshore region off a river of a large continent vs. nearshore region off a small desert island

_____

_____

5. In the Antarctic and equatorial western Pacific, nutrients are supplied throughout the year by upwelling mechanisms. Thus, light availability is the major factor controlling primary productivity. Considering the seasonal changes in light availability in each region, plot the hypothesized relative intensity of primary productivity for each location. Explain your reasoning in the space below.

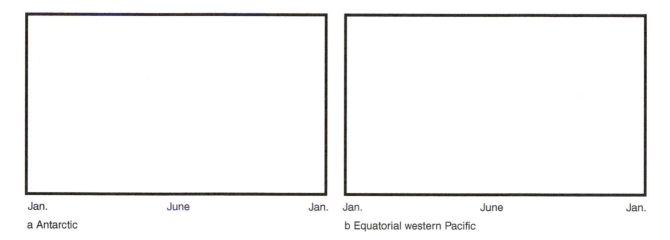

Jan.   June   Jan.   Jan.   June   Jan.

a Antarctic   b Equatorial western Pacific

6. Examine the distribution of surface primary productivity in the world ocean shown in Color Plate 8. Note the color scale that indicates the relative abundance of phytoplankton in surface waters. For each letter on the figure, explain the observed level of productivity using appropriate oceanographic mechanisms (e.g., upwelling, geostrophic currents, river runoff, etc.).

(A) North Pacific: _____

_____

(B) California Current: _____

_____

(C) Equator: _____

_____

(D) Southern Ocean: _____

_____

(E) Gulf Coast: _____

_____

7. Below is a graph of relative seasonal variations in light availability, temperature, nutrients, and plankton standing crops. Starting from the left end of the graph, examine how the different variables increase and decrease relative to one another over a year. Using these relationships, answer the following questions:

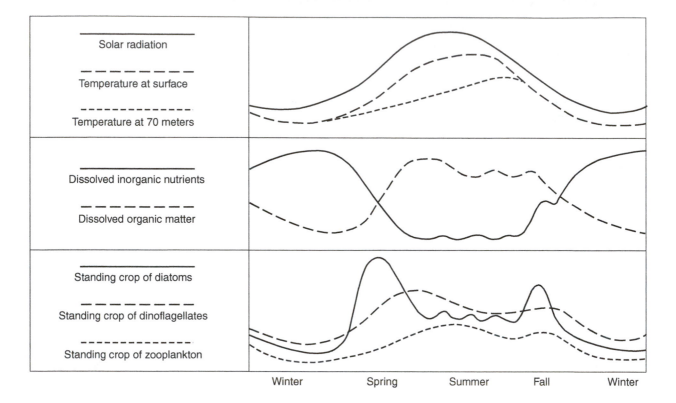

(a) When are surface and 70-meter temperatures most similar? _____
Most different? _____
What causes the two to diverge and converge over the year? _____
_____
_____

(b) Why do dissolved inorganic nutrients decrease in the spring? _____
_____

(c) Why does the diatom "bloom" (rapid increase) start in the spring and end in the early summer?
_____

(d) What process(es) may explain the different rates of change in the diatom and dinoflagellate standing crop curves during the annual cycle? _____
_____

(e) Why is the zooplankton standing crop always less than that of the other plankton? _____

(f) What are some likely physical mechanisms for the nutrient increase in the late fall? (Hint: Consider the surface and 70-meter temperatures during this time.) _____
_____

(g) As nutrient concentrations increase during the late fall, diatoms "bloom" again, but then rapidly drop, although nutrient concentration continues to increase. Explain this pattern. _____
_____

El Niño

OBJECTIVES:

■ To know the atmospheric and oceanographic factors leading to El Niño/Southern Oscillation (ENSO) "events."

■ To understand the impact of El Niño on humans and on marine life.

Although it may be hard to believe from the common references in broadcast and print media, before the winter of 1997–98, only oceanographers and Pacific fishermen were familiar with El Niño. This oceanographic phenomenon in the Pacific Ocean was originally noticed by fishermen along the Pacific coast of Peru and Ecuador, when periodic appearances of warm, nutrient-poor ocean waters led to decreases in the local fishing industry. In 1892, fishermen in the Peruvian port of Paita called an invasion of warm water off the coast around Christmas time "Corriente del Niño" (current of the Christ Child). In 1958, the term *El Niño* was proposed for the oceanographic and meteorological events that occurred across the Pacific Ocean during December of that year, and persisted for several months before the restoration of "normal" ocean currents and surface-water temperatures. In 1972, oceanographers and meteorologists used the term **El Niño-Southern Oscillation (ENSO),** broadening the definition of these events to include periodic climate-related shifts in weather patterns in the equatorial Pacific and Indian oceans. To complicate matters, during the 1990s, the news media began to refer to times when anomalously *cold* surface-water masses appeared in the eastern equatorial Pacific Ocean as *La Niña* events.

The Generation of El Niño Events

Occurrences of unusually warm surface water in the Pacific Ocean are episodic, and eight severe El Niño events have occurred since the late 1950s. Although the 1997–98 El Niño was widely reported in news broadcasts, it remains to be proved whether its effects exceeded those of the 1982–83 El Niño, which was the strongest event of the twentieth century, affecting the entire Pacific Rim and the Indian Ocean.

To understand the El Niño-Southern Oscillation, it is first necessary to understand the "normal" circulation of the equatorial Pacific Ocean (Figure 17-1). During normal times, the coastal current along the Pacific margin of South America is the Peru Current (also previously known as the Humboldt Current), which moves cold waters northward from Antarctic regions. Southeast trade winds in the equatorial Pacific Ocean, combined with Coriolis deflection to the left in the Southern Hemisphere, push the water in the northward-flowing Peru Current offshore, causing surface waters to be replaced by upwelling of cool, nutrient-rich water from below the thermocline (Figure 17-2). This cold upwelled inshore water is so distinct that it is called the *Peru Coastal Current* to distinguish it from the offshore Peru Current. The Peru Coastal Current supports one of the richest fisheries in the world, and commercial fishermen from southern California travel thousands of miles to the coastal waters off Peru.

Under normal conditions, the North Equatorial Current and the South Equatorial Current force surface waters westward across the Pacific. These surface waters are warmed in their transit across

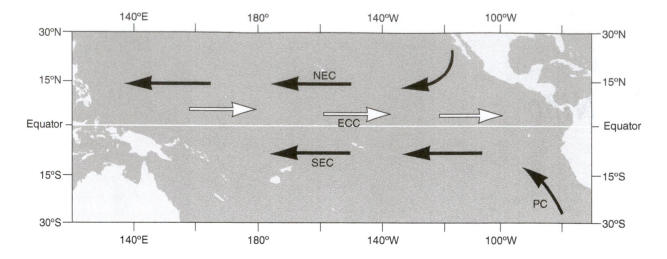

Figure 17-1 Surface currents in the equatorial Pacific Ocean. NEC = North Equatorial Current; SEC = South Equatorial Current; ECC = Equatorial Countercurrent; PC = Peru Current.

the ocean, forming a distinct "bulge" of warmer water in the western equatorial Pacific, which may be seen on satellite images showing relative sea level. During El Niño years, the usual low atmospheric pressure in the western equatorial Pacific is replaced by high-pressure cells. This shift in pressure is known as the Southern Oscillation. Since this change lessens the pressure differential that drives the trade winds from east to west, those winds decrease or even reverse direction, allowing the huge mass of warm surface water in the western Pacific to surge back toward South America in an intensified Equatorial Countercurrent (see Figure 17-1).

When this warm, dilute, nutrient-poor water strikes the coast of South America, it moves southward over the cooler coastal waters, forming a "lid" that stops the upwelling of the Peru Coastal Current. Lower nutrient concentrations in surface waters result in the disappearance of fish species that depend on the upwelling of cool, nutrient-rich water, and thus the collapse of the local fishing industry. The disappearance of these fish also leads to the deaths of millions of seabirds that normally feed on them. In turn, the local guano fertilizer industry vanishes because of the decrease in seabird abundances. Eventually, coastal communities in Peru and Ecuador experience economic hardship and sometimes even famine.

Monitoring El Niño

The onset of El Niño events can be predicted by oceanographic changes in the central equatorial Pacific Ocean before they affect the west coasts of North and South America. NOAA has attempted to track and predict the onset and relative impact of El Niño events. The NOAA Pacific Marine Environmental Laboratory maintains the **Tropical Atmospheric Ocean Array** (TAO), an array of 70 buoys moored in the central Pacific, between 160°E and 150°W and from 5°S to 5°N, in an area termed the Nino-4 Region (Figure 17-3). NOAA-TAO buoys measure temperature, winds, and currents in the Nino-4 region, and transmit these data back to the Pacific Marine Environmental Laboratory.

One means of tracking El Niño is by monitoring sea-surface temperatures. Under normal conditions, surface waters in the eastern Pacific are cooler (around 22°C) than those in the western Pacific (around 30°C). If temperatures in the eastern Pacific are *warmer* than usual, this is a positive SST anomaly. Conversely, negative SST anomalies mark times when surface waters there are *cooler* than usual. El Niño events may be seen in SST data collected by the TAO array (Color Plate 6). During El Niño, higher than usual temperatures displace the typical "cold tongue" of water that normally

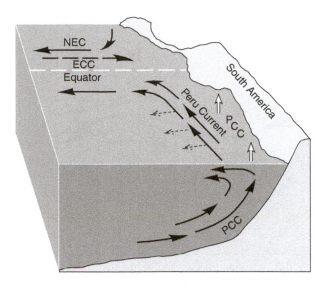

Figure 17-2 Three-dimensional view of upwelling off the western coast of South America. The cool, nutrient-rich upwelled waters of the Peru Coastal Current (PCC) create a rich fishing ground off Peru and Ecuador. Cessation of upwelling during El Niño events leads to the collapse of the fishery in this area. NEC = North Equatorial Current; SEC = South Equatorial Current; ECC = Equatorial Countercurrent.

extends westward across the equatorial Pacific Ocean. The atmospheric pressure shifts that trigger El Niño are also periodic, and can be detected by orbiting weather satellites.

Effects of El Niño

The El Niño-Southern Oscillation can cause global shifts in weather patterns, as well as latitudinal changes in the position of the jet streams (stratospheric winds that affect weather patterns by "steering" high- and low-pressure cells, and thus weather fronts). In particular, deflection of the subtropical jet stream (often called the "Pineapple Express" because it flows near Hawaii) to the north may allow tropical storms to flow northward, causing flooding as far north as Washington and Oregon (Figure 17-4). The 1982–83 ENSO was responsible for flooding and droughts in 12 countries, thousands of deaths, and billions of dollars in property damage. In southern California, the rainfall nearly tripled, and the coast was battered by destructive winds and high waves.

An example of the effects of El Niño on southern California is shown by the Advanced Very High Resolution Radiometer (AVHRR) image in Color Plate 7. These color-coded pictures show the warmest sea-surface temperatures in red and orange, intermediate-temperature areas in yellow and green, and the coldest regions in blue; clouds appear white. The left-hand photo (taken in January 1982) depicts normal conditions, with cool California Current water flowing southward past Point Conception (which corresponds to the area labeled 1 in the photo on the right). The cool coastal waters (green) are typical of southern

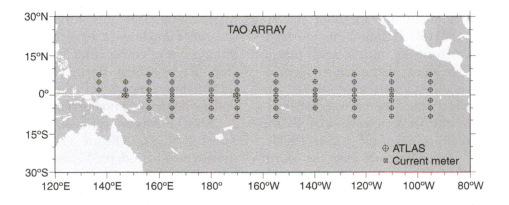

Figure 17-3 The Tropical Atmospheric Ocean Array (TAO) of oceanographic sensor buoys and moored current meters in the Niño-4 region of central Pacific Ocean (between 160°E and 150°W, and from 5°S to 5°N). Data obtained from these sensors is used by NOAA to monitor oceanic conditions that might mark the onset of El Niño events.

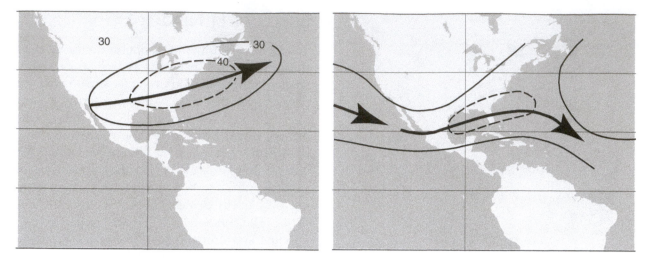

"Normal" jet stream
December – February

El Niño jet stream
December 1982 – February 1983

Figure 17-4 Typical December–February location of the subtropical jet stream (*left*) and northward deflection of the subtropical jet stream during the El Niño of 1982–83 (*right*). [After Eugene M. Rasmusson, "El Niño: The Ocean/Atmosphere Connection." *Oceanus,* Vol. 27, no. 2, 1984, pp. 5–13].

California and Mexico, where winds from the north cause upwelling along the coast. These cool, nutrient-enriched coastal waters give rise to plankton blooms and high biological productivity.

The right-hand photo, which was taken in January 1983, shows the effects of El Niño. South of Point Conception, surface waters are 2°C warmer than normal. Abnormal onshore winds have caused warm waters to pile up along the coast, raising the average sea level 20 centimeters above normal. In addition to temperature and sea-level effects, the 1982–83 El Niño brought with it exotic species of marine life, such as small pelagic red crabs and large squid, that are rarely seen in southern California coastal waters.

During El Niño years, the presence of high-pressure cells in the western equatorial Pacific disrupts the rainfall regime of the tropical Pacific. Normally, heavy rainfall occurs over the North Australian-Indonesian region and along the South Pacific convergence zone from New Guinea eastward to the International Date Line. The 1982–83 El Niño was marked by severe droughts in Australia, Indonesia, and the western equatorial Pacific. It was also associated with dust storms and brush fires in Australia. Drought during the 1997–98 El Niño caused forest fires of such

severity that aircraft were unable to land at Indonesian airports because of impenetrable smoke.

El Niño events may also affect the Atlantic Ocean, as the northward shift of the subtropical jet stream allows warmer, moister air to flow eastward into the Gulf of Mexico and the Caribbean Sea. The invasion of this air from the Pacific tends to slow down the development of tropical storms and hurricanes in the Atlantic basin. This is because Atlantic tropical storms increase in strength and can evolve into hurricanes as they move northward into regions of cooler, drier air at higher latitudes. Only seven named storms formed during the Atlantic hurricane season of 1997, and only three of these developed into hurricanes.

ENSO events are periodic, with a return interval of 3–7 years (Figure 17-5). Because of this short return period, predicting these events is important to government preparedness agencies and the general public. Pacific equatorial sea-surface temperatures determined by the TAO buoy network are shown in Figure 17-5. Averaging temperatures over this region allows oceanographers to identify positive (warmer) temperature anomalies that may signal the onset of an ENSO event. Pronounced events may last for an entire year be-

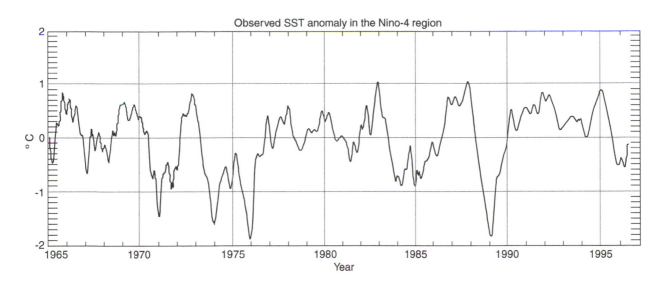

Observed SST anomaly in the Nino-4 region

Figure 17-5 The periodicity of El Niño is demonstrated by the sea-surface temperature anomalies observed over time in the Pacific Ocean near 180° latitude at the equator. Positive anomalies are temperatures higher than average, and mark the warm surface waters of El Niño events.

fore conditions return to normal, but most events persist for only 3–4 months. La Niña events are seen as negative temperature anomalies; that is, cooler than usual surface waters move eastward across the equatorial Pacific Ocean.

How do El Niño events end? The accumulation of warm surface waters in the eastern Pacific causes a weakening of the westward-flowing trade winds, leading to the further strengthening of ENSO events by positive feedback. However, sea-surface height in the western Pacific eventually decreases as surface waters "slosh" across the equatorial Pacific in eastward-flowing *Kelvin waves*. Kelvin waves are similar to seiches (see Exercise 10), having an up-down motion similar to that of the wave that develops in a bathtub as the bather moves, causing water to slosh back and forth in the tub. These slow-moving waves eventually slosh back to the eastern Pacific, leading to the reestablishment of the normal east-to-west motion of surface currents. As the surface currents are restored, the eastern equatorial Pacific cools, and the pole-to-equator motion of the Peru Current restores the normal upwelling regime of the Peru Coastal Current.

Web Sites

Current information on sea-surface temperatures, El Niño, and La Niña, with animations and inter-

active questions and answers, may be found at the following sites:

http://www.elnino.noaa.gov/lanina.html
NOAA site

http://www.ogp.noaa.gov/enso/   More NOAA

http://www.pmel.noaa.gov/toga-tao/ la-nina-story-nas.html   NOAA and TAO Buoy array

http://www.pmel.noaa.gov/toga-tao/ el-nino/home.html   with animations of SST in the Pacific

http://www.pbs.org/nova/elnino   Public Broadcasting System *Nova*

http://www.nws.noaa.gov   National Weather Service

http://observe.ivy.nasa.gov/nasa/earth/ el nino/elnino1.html   NASA's site

DEFINITIONS

El Niño.   An oceanographic event occurring on a repetitive basis in the equatorial Pacific Ocean, marked by abnormally warm surface waters and reduced upwelling in the eastern Pacific.

Kelvin wave. An ocean wave with horizontal surging, like water sloshing back-and-forth in a basin, instead of the more familiar orbital motion of nearshore waves.

La Niña. Repetitive oceanographic events in the equatorial Pacific Ocean, marked by colder than usual surface waters in the eastern Pacific.

Peru Current (also known as the Humboldt Current). Surface current flowing along the Pacific margin of South America, moving cold waters northward.

Peru Coastal Current. Upwelled colder waters, occurring inshore from the Peru Current. The Peru Coastal Current supports a rich commercial fishery, but slows or even stops entirely, during El Niño events.

TAO Array. The Tropical Atmospheric Ocean array, seventy buoys moored in the central Pacific Ocean, used by NOAA to monitor surface water temperature, current speed and direction, and wind speed and direction.

Exercise 17

El Niño

NAME _____

DATE _____

INSTRUCTOR _____

1. Use the information in Figure 17.5 to determine the periodicity of El Niño.
   (a) How many El Niño events can be recognized in this figure? (Keep in mind that a single event may last more than one year.) _____
   (b) How many years elapse between El Niño events? _____
   (c) How many La Niña events (cooler than usual conditions) can be recognized? _____
   (d) How many years elapse between these events? _____
   (e) Is the periodicity of El Niño events the same as that of La Niña events? _____

2. Compare the satellite photos of El Niño in Color Plate 7 to the map of the California Current in Figure 9-9.
   (a) Based on the data presented in both of these figures, sketch the direction of flow of El Niño along the California coast on the outline map below.

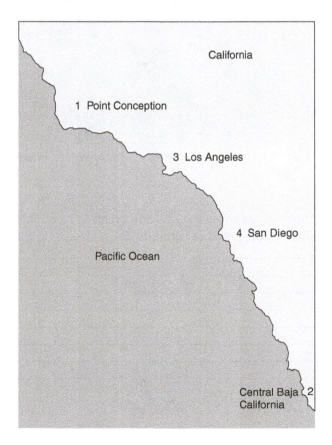

(b) In a few sentences, describe the flow of El Niño as compared to normal current flow (use Figure 9-9 for reference if needed). _____

_____

_____

_____

_____

_____

_____

_____

3. On the outline map in Question 2, draw arrows showing the wind directions that produce upwelling (normal conditions) along this stretch of coastline. Using dashed lines and arrowheads, show the wind directions that caused destruction and severe beach erosion along the coast during the ENSO of 1982–83.

4. The figure below shows the locations of selected oceanographic buoys from the NOAA-TAO array in the central equatorial Pacific, with mean sea-surface temperature (SST) anomaly data indicated for each buoy. Contour these data, using 0.5° intervals.

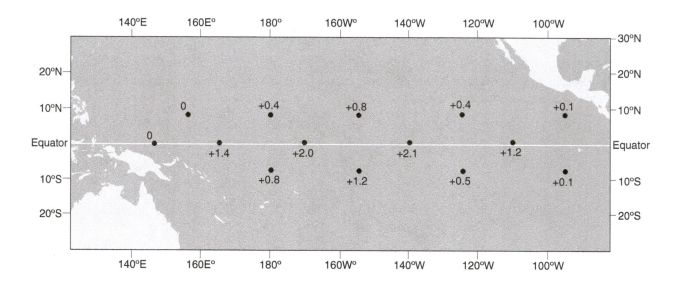

(a) Do these data indicate normal conditions, an El Niño event, or La Niña? Justify your answer by comparing to the SST anomaly data plotted in Figure 17-5. _____

_____

_____

_____

_____

_____

_____

_____

5. The figure below shows SST anomaly data for the same buoys as in Question 4, but from a different time period. Once again, contour these data, using 0.5° intervals.

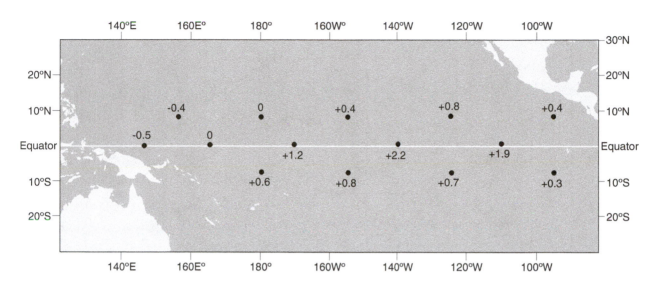

(a) Compare these data to those in Question 4. Is there anything you might predict for future climatic conditions on the west coast of South America, based on these data? _____
_____
_____
_____
_____

(b) If these data indicate an event, which type? _____
_____
_____
_____
_____

(c) Assuming a time difference of one month between these data and those in Question 4, when might climatic conditions change in Guayaquil, Ecuador (3°S, 82°W)? _____
_____
_____
_____
_____

6. Why would oceanographers in the Atlantic Ocean and the Gulf of Mexico be concerned about a La Niña event? _____
_____
_____
_____
_____

What effect would La Niña have on the Atlantic hurricane season?   _____
_____
_____
_____
_____

Marine Ecosystems and Nutrient Cycles

OBJECTIVES:

■ To understand the role of the ocean as an ecosystem and nutrient recycler.

■ To understand the interactions between and flow of energy through producers, consumers, and decomposers.

■ To appreciate how humans can disrupt marine ecosystems.

In this exercise we explore how energy from primary productivity (Exercise 16) is transferred through an ecosystem. An ecosystem consists of a group of living organisms, the physical environment in which they live, and an energy source (e.g., sunlight in photosynthesis-based ecosystems). The largest ecosystem can be considered the earth as a whole; the planet may be subdivided into terrestrial and marine ecosystems, and each of these may be further subdivided, often on the basis of environmental conditions (e.g., depth, temperature, etc.). In each ecosystem, there are organisms that produce food (primary producers or **autotrophs**), organisms that consume other organisms (secondary producers or **heterotrophs**), and organisms that decompose autotroph and heterotroph waste products and bodies after death (**decomposers;** generally fungi and bacteria). Ecosystems have two fundamental properties (Figure 18-1). The first is that energy flows through an ecosystem in only one direction: it is received from the sun, transformed into organically usable forms through primary producers, and flows to secondary producers and decomposers. Depending upon the complexity of the ecosystem, this one-way transfer of energy may be represented by a **trophic pyramid** (Figure 18-2) or **trophic web** (Figures 18-3 and 18-4). The second property is that individual nutrients, elements necessary for metabolic processes, are recycled many times within most ecosystems, with decomposers playing a crucial role in the release of nutrients from organic matter.

Trophic Pyramids and Webs: Examples from the Antarctic Ocean

A simplified trophic pyramid for the Antarctic Ocean is presented in Figure 18-2. Diatoms are the primary producers, providing energy for the entire ecosystem, and are shown at the base of the pyramid. These primary producers are consumed by the primary consumers (herbivores) of the trophic pyramid's second tier, mostly krill. In turn, krill are the energy source for the third trophic tier, the whales. The whales are termed secondary consumers (they eat primary consumers) or first-level carnivores (the first to eat heterotrophs).

In Figure 18-3, a more realistic model of energy flow through the Antarctic Ocean ecosystem is presented. This trophic web is essentially a trophic pyramid expanded to include the more complex interrelationships between organisms at higher trophic levels; the trophic relationship between the diatoms (primary producers) and krill (secondary consumers or herbivores) remains the same. Note that there is a greater diversity of organisms at the higher trophic levels and that some of these can operate at multiple trophic levels. In this trophic web, blue whales remain in the third trophic level,

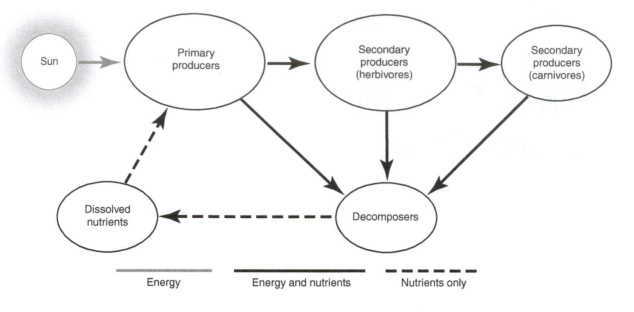

Energy   Energy and nutrients   Nutrients only

Figure 18-1 Generalized ecosystem diagram showing the one-way flow of energy and the recycling of nutrients. Note that nutrient recycling is not "perfect"; some organic matter sinks out of the photic zone before it is broken down by decomposers, and some fraction of this amount is buried (not shown).

but other organisms directly dependent upon krill are also included (i.e., crabeater seals, winged birds, Adélie penguins, and small fish and squid). The small fish and squid, in turn, are prey items for emperor penguins, larger fish, and Weddell and Ross seals—members of the fourth trophic level. Because skuas feed upon the chicks of Adélie penguins, they are also considered members of the fourth trophic level. The remaining organisms in the trophic web, the leopard seals and killer whales, occupy multiple trophic levels, as they feed upon multiple trophic levels below them. For example, leopard seals may be assigned to the fourth trophic level when feeding upon crabeater seals or Adélie penguins, or the fifth trophic level when feeding upon emperor penguins, large fish, or Weddell and Ross seals. Similarly, the killer whale belongs to the fourth trophic level when feeding upon blue whales or crabeater seals, or to the fifth trophic level when feeding upon leopard seals. Note that some organisms may shift their trophic position during their lifetime; for example, as fish grow larger, some shift from the second to third trophic level. Appreciate that 13 different species are directly dependent upon krill, which are themselves dependent upon diatoms.

Compare the Antarctic food web to that of the Long Island estuary (Figure 18-4). Notice that trophic levels in the estuary increase from left to right, from primary producers (plants, phytoplankton) to top carnivores (birds). Such complexity is characteristic of many marine food webs, with many interrelations between organisms. Ecological theory states that the greater the number of food pathways leading from the primary producers to higher trophic levels, the more resistant the ecosystem will be to disturbances related to losses of individual species. Why might this be so?

Ecological Efficiency and Biological Magnification

Energy contained within an ecosystem is not recycled, but moves unidirectionally through successively higher trophic levels. Within a given trophic level, the vast majority of energy obtained from the level below is used for respiration and metabolism and is lost as excretion or heat; only a small portion is transformed into biomass through growth. In addition, not all available individuals in the

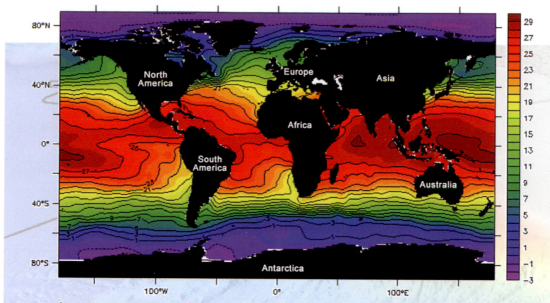

Annual mean ocean surface temperature in degrees Celsius (red = high; purple = low)

Color Plate 1 Global sea-surface temperatures (SST), based on shipborne and buoy measurements. Warm-water temperatures (greater than 25 degrees C, or 77 degrees F) are in red, intermediate temperatures are in yellow, and cooler water (below 5 degrees C, or 41 degrees F) is shown in blue. *[NOAA.]*

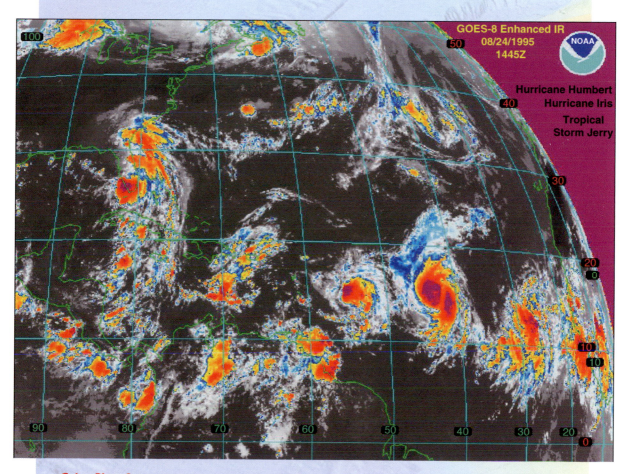

Color Plate 2 Geostationary Operational Environmental Satellite (GOES) image taken August 24, 1995, clearly shows the development of tropical waves into tropical storms and hurricanes in the equatorial Atlantic Ocean. *[Courtesy of NOAA, National Climatic Data Center.]*

Color Plate 3 Every day the Moderate-resolution Imaging Spectroradiometer (MODIS) measures sea surface temperature over the entire globe with high accuracy. This false-color image shows a one-month composite for May 2001. Red and yellow indicates warmer temperatures; green is an intermediate value; blues and then purples are progressively colder values. *[Courtesy of NASA.]*

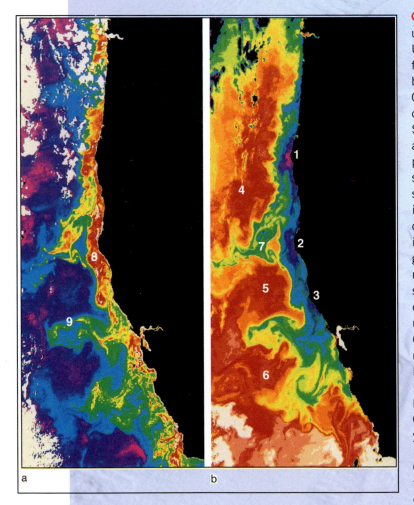

a   b

Color Plate 4 Two images of upwelling on the west coast of the United States taken 8 hours apart from NASA's Nimbus-7 satellite. The black area is coastal Oregon and California. (a) Biological productivity derived from the Coastal Zone Color Scanner (CZCS). As phytoplankton and their associated chlorophyll pigments increase, the water color shifts from blue to green. The CZCS sensor detects this color shift, which is used to calculate chlorophyll concentrations and render the color image. Chlorophyll concentrations greater than 10 milligrams/cubic meter are shown in red (8). (b) Sea-surface temperature (SST) readings derived from the AVHRR. Freshly upwelled cold water in violet-purples (8–9°C or 47–49°F) is especially noticeable at Cape Blanco (1), Cape Mendocino (2), and Point Arena (3). Intermediate temperatures are shown by green and blue, and warmer California Current water is shown in yellow and red (14–15°C or 57–59°F). Meanders in the California Current are seen at 4, 5, and 6, and filaments of upwelled water (7) extend 100–300 kilometers offshore. *[Courtesy of NASA.]*

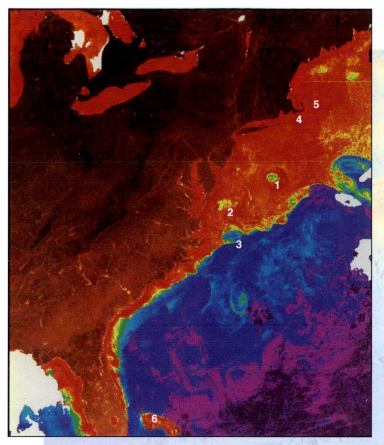

Color Plate 5 Chlorophyll concentrations in the western North Atlantic Ocean. When phytoplankton are abundant, their chlorophyll pigments absorb blue light, making the water appear greener. The CZCS images the ocean in four narrow bands, and the changes in the ratio of green to blue light are used to calculate chlorophyll concentrations. This image is a composite of eight satellite passes to minimize areas with cloud cover and gain maximum aerial coverage. The highest chlorophyll content (1 gram/cubic meter) is represented by red-brown; concentrations decrease through orange, yellow, green, and blue, with the lowest (< 0.01 gram/liter) represented by purple. *[Courtesy of NASA.]*

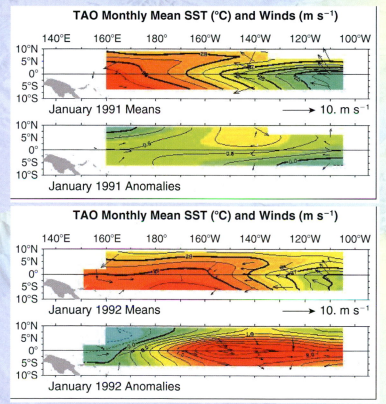

Color plate 6 Monthly mean sea-surface temperature data for the Niño-4 region, compiled by the Tropical Atmospheric Ocean Array (TAO) of the NOAA Pacific Marine Environmental Laboratory. January, 1991 is a normal (non-El Niño) year, while January, 1992 shows data for an El Niño event. *[Courtesy of NOAA.]*

Color Plate 7 Advanced Very High-Resolution Radiometer (AVHRR) satellite image of southern California, showing normal conditions during January, 1982 *(left)* and El Niño conditions during January 1983 *(right)*. *[Photo by Paul Fiedler, National Marine Fisheries Services/Southwest Fisheries Center, processed at Scripps Satellite Oceanography Facility.]*

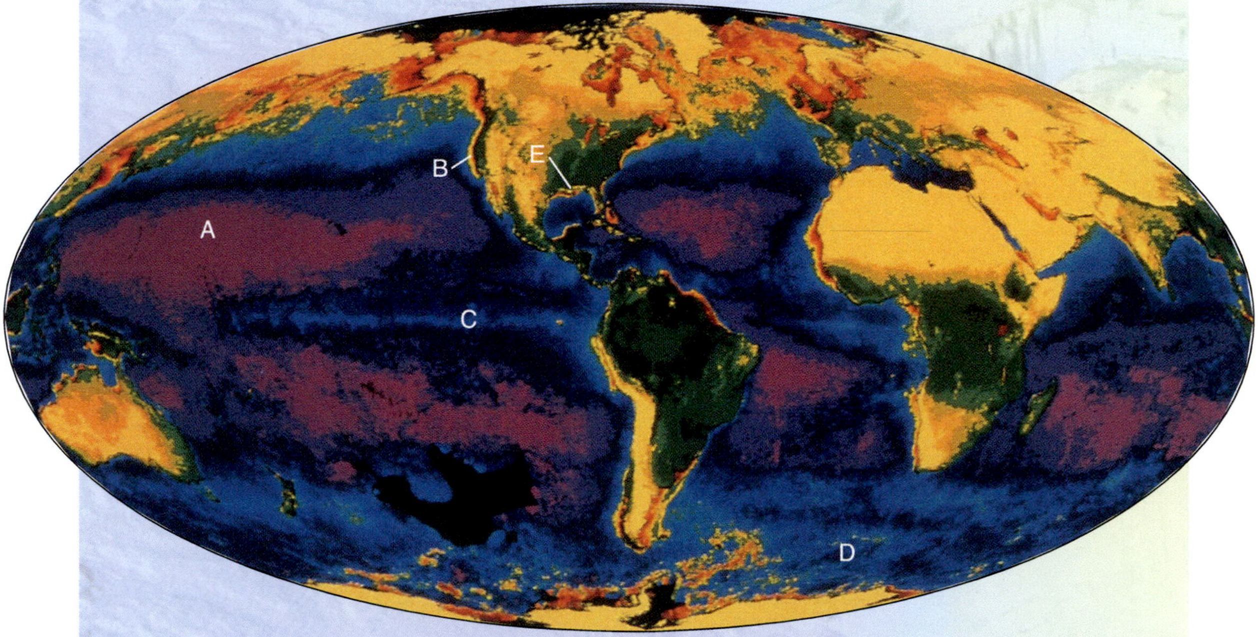

Color Plate 8 A view of phytoplankton productivity in the world's oceans. Bright blue indicates high productivity and therefore chlorophyll content. *[Courtesy of NASA.]*

lower trophic level will be consumed; some will die of natural causes. Both of these factors result in a phenomenon termed **ecological efficiency,** which expresses the amount of energy (biomass) flowing into a trophic level compared to the amount of energy retained within that trophic level. In most ecosystems, ecological efficiency is less than 10 percent. For example, in Figure 18-2, the numbers on the left indicate that of the roughly 1000 biomass units of the first trophic level, only 100 units (∼ 10 percent) of this energy is converted into biomass at the second trophic level. Another way to visualize this relationship is that 100 grams of diatoms are required to support 10 grams of krill, and 10 grams of krill are required to support 1 gram of blue whale. From a fishery standpoint, the greater the number of trophic levels between primary producers and edible fish, the less efficient the ecosystem is in converting solar energy into food products for humans.

Related to this concept of ecological efficiency is the biological concentration of pesticides, such as DDT, in the marine ecosystem. DDT is an extremely effective agent against agriculture-damaging insects and has been in extensive use since the 1940s, although banned in the United States since 1972. One reason the pesticide is so effective is its high resistance to biological breakdown. This resistance leads to DDT's eventual transport from agricultural fields through irrigation and runoff into the ocean, where it is incorporated into the marine biosphere during primary production. Like most pesticides, DDT is damaging to non-insects as well; in phytoplankton, it decreases the efficiency of photosynthesis and therefore reduces the total primary production supporting higher trophic levels.

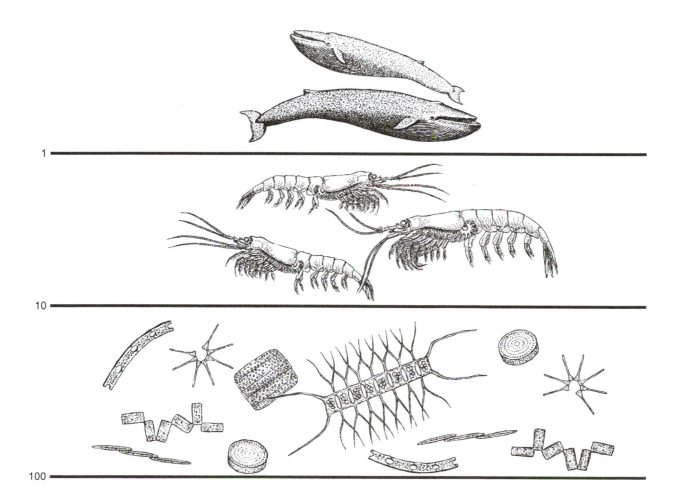

Figure 18-2 A simplified trophic pyramid for the Antarctic ocean. [After Robert C. Murphy, "The Oceanic Life of the Antarctic." Copyright © 1962 by Scientific American, Inc. All rights reserved.]

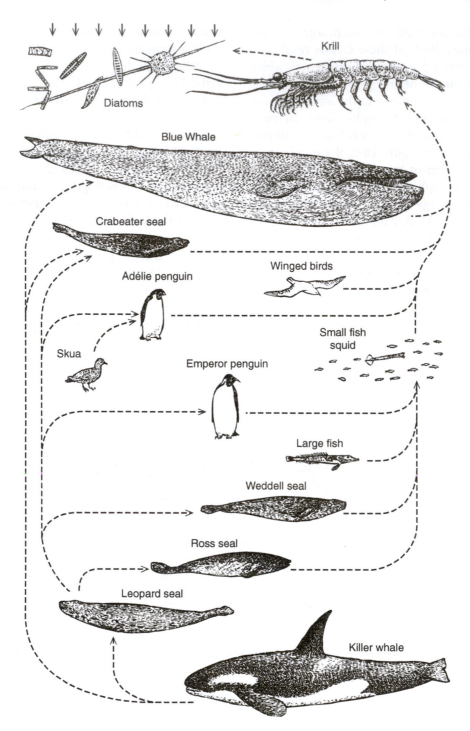

***Figure* 18-3** Summary of major trophic relationships within the Antarctic ecosystem. [After Robert C. Murphy, "The Oceanic Life of the Antarctic." Copyright © 1962 by Scientific American, Inc. All rights reserved.]

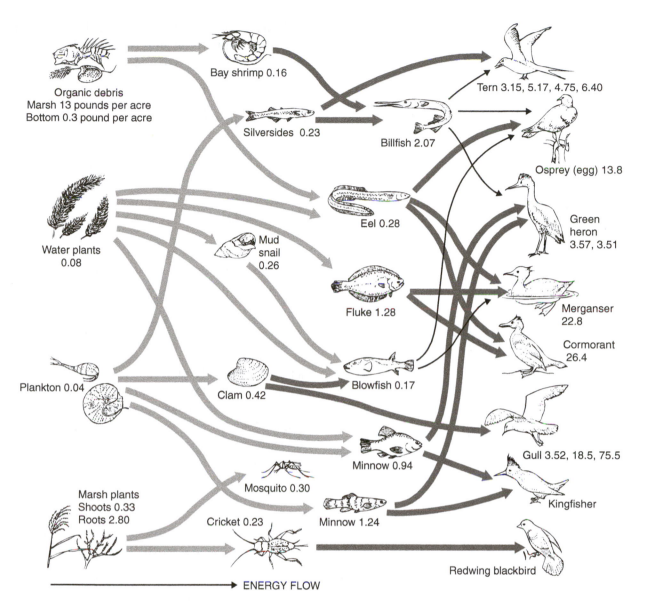

Organic debris
Marsh 13 pounds per acre
Bottom 0.3 pound per acre

Bay shrimp 0.16

Silversides 0.23

Billfish 2.07

Tern 3.15, 5.17, 4.75, 6.40

Osprey (egg) 13.8

Water plants
0.08

Mud snail 0.26

Eel 0.28

Green heron 3.57, 3.51

Plankton 0.04

Clam 0.42

Fluke 1.28

Blowfish 0.17

Merganser 22.8

Cormorant 26.4

Minnow 0.94

Gull 3.52, 18.5, 75.5

Marsh plants
Shoots 0.33
Roots 2.80

Mosquito 0.30

Cricket 0.23

Minnow 1.24

Kingfisher

Redwing blackbird

ENERGY FLOW

Figure 18-4 Summary of major trophic relationships within the Long Island estuary. The numbers beside each organism's name refer to the average DDT content, in parts per million, found in each organism. [After George M. Woodwell, "Toxic Substances and Ecological Cycles." Copyright © 1967 by Scientific American, Inc. All rights reserved.]

Unfortunately, the effects of DDT on the marine ecosystem do not end at the first trophic level. Because it is not easily broken down, the pesticide is stored in body tissue and transferred up through higher trophic levels. When DDT-enriched phytoplankton are consumed by primary consumers, the 10 percent efficiency effect applies to organic compounds, but not to DDT. Thus, the DDT is passively transferred through the ecosystem. While some DDT is given off through respiration and excretion at each level, the bulk remains in the bio-

mass and becomes increasingly concentrated at higher trophic levels, becoming the highest in the top carnivores (Figure 18-5).

This phenomenon is called *biological magnification,* and the resulting DDT concentrations can wreak havoc with biological processes ranging from eggshell calcification in birds to reproduction in humans. For example, in the final years before the 1972 U.S. ban on DDT, brown pelicans, which nest on islands off the coast of southern California, had acquired so much DDT that most of their

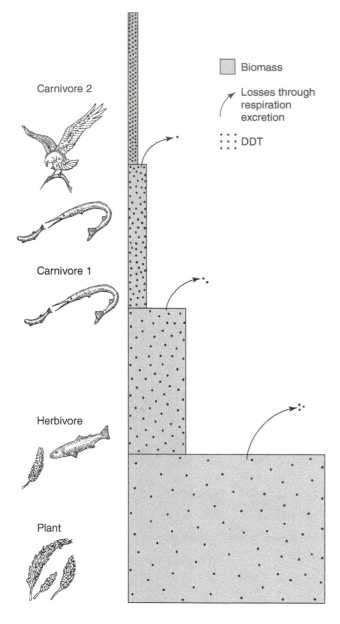

Figure 18-5 Biological magnification of DDT in a marine food chain. [After George M. Woodwell, "Toxic Substances and Ecological Cycles." Copyright © 1967 by Scientific American, Inc. All rights reserved.]

eggshells would break when incubated, causing the pelican population to drop precipitately. Since the 1972 ban, the pelican population has recovered, but DDT concentrations remain relatively high. Concentrated levels of DDT have also been found in Adélie penguins and skuas, which indicates that the pesticide has been spread through oceanic circulation and other processes to regions far removed from its original application. You most

likely have some amount of DDT in your body right now, and it will only increase during your lifetime. Good news? No. A fact? Yes.

Biogeochemical Cycles

The main factor limiting productivity in well-lit marine waters is the availability of principal **nutrients** such as nitrate, phosphate, and carbon (although carbon is not limiting in most regions). Here we will examine the cycling of these nutrients through various "reservoirs" in the atmosphere, lithosphere, hydrosphere, and biosphere, and see how their distribution and exchange affects the marine ecosystem. The general biogeochemical cycle involves the intake of an inorganic form of a nutrient by an autotroph during photosynthesis of organic molecules, which are subsequently transferred through the trophic web by heterotrophic activity. Eventually, decomposers transform dead organic matter and waste back into its inorganic form, making the constituent nutrients available again to autotrophs. Keep in mind that decomposers, such as bacteria and fungi, are critical components in recycling such nutrients.

Carbon Cycle. Carbon is the basic building block of all organic molecules. Its biogeochemical cycle is presented in Figure 18-6. Note that carbon is rarely limiting in most marine ecosystems; only about 1 percent of the carbon reservoir in the ocean is involved in primary productivity at any given time. Another major carbon reservoir is atmospheric carbon dioxide (CO_2), which enters the ocean by gas exchange at the air-water interface. Additional carbon dioxide comes from the respiration of plants and animals. Thus, dissolved carbon dioxide is readily available to autotrophs for photosynthesis, which "fixes" carbon into organic molecules that are then transferred through the trophic web by predation and feeding. Carbon is also present in carbonate sediments and limestones composed of the calcareous skeletons of marine organisms, and as natural petroleum deposits (i.e., oil, gas, peat, coal) formed by burial of ancient organic material. Carbon within these reservoirs tends to have a long residence time (the average time interval that individual particles remain in a given reservoir), but is ultimately cycled through terrestrial weathering in dissolved form to the

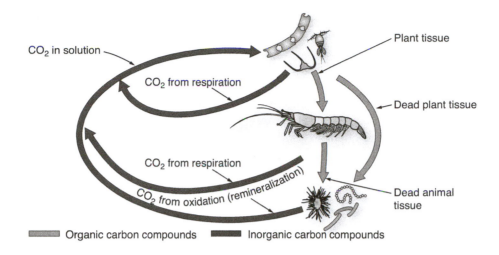

Figure 18-6 The biogeochemical cycle of carbon in organic and inorganic forms.

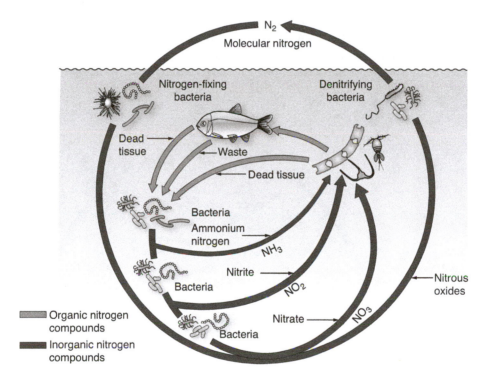

Figure 18-7 The biogeochemical cycle of nitrogen in organic and inorganic forms.

ocean or oxidized by natural or human-related burning into the atmosphere.

Nitrogen Cycle. Nitrogen is a essential element in the production of amino acids—the building blocks of proteins within all living things. The biogeochemical cycle of nitrogen is summarized in Figure 18-7 and involves a fairly complex suite of bacterial "fixers" and decomposers. Molecular nitrogen (N_2) cannot be used directly by most organisms, but metabolic processes within cyanobacteria convert dissolved nitrogen into nitrate (NO_3) through a process called *nitrogen fixation.* The resulting nitrate is the nutrient form most easily utilized by most phytoplankton and is subsequently transferred up the trophic web through feeding.

Excreted waste and dead organic matter are broken down by decomposers as an energy source. Some of these decomposers are *denitrifying* bacteria, whose metabolism breaks down organic-bound nitrogen into progressively oxidized inorganic forms: the first is ammonia (NH_3), followed by nitrite (NO_2), or finally nitrate (NO_3). The less oxidized forms (ammonia, nitrite) are generally taken up again by autotrophs before denitrifying bacteria can fully oxidize the compounds to nitrate. Only a small percentage of nitrogen initially fixed by cyanobacteria is recycled within the photic zone; the majority is not oxidized back into a usable form until it is well below the photic zone and thus unavailable to photosynthesizing autotrophs. The major mechanisms for returning this fixed nitrogen to the photic zone are seasonal vertical mixing of the water column and local upwelling of nutrient-rich bottom waters. Some denitrifying bacteria will completely oxidize the nitrogen-bearing compounds back into molecular nitrogen, some of which may be exchanged with the atmosphere. In addition, small amounts of nitrogen are buried in ocean sediments and released through terrestrial weathering—for simplicity and because of their relatively small contribution to the marine cycle, these paths are not shown in Figure 18-7.

Phosphorus Cycle. Phosphorus is an essential element in all living organisms' genetic information (DNA and RNA) and in the ATP compounds involved in the conversion of carbohydrates into energy. Its biogeochemical cycle is simpler than that of nitrogen, largely because the bacterial component of the cycle is simpler (Figure 18-8). Phosphorus is released to the ocean through the weathering of phosphate-bearing rock and removed through burial of organic matter (not shown). Inorganic orthophosphate may be used directly by autotrophs and, like nitrogen, is transferred to higher trophic levels through feeding. The phosphorus within excreted waste and dead or-

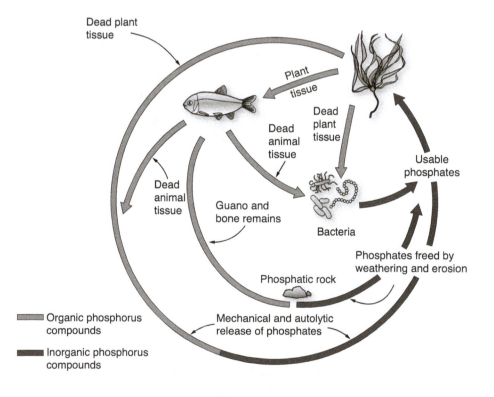

Figure **18-8** The biogeochemical cycle of phosphorus in organic and inorganic forms.

ganic matter is released back into the environment through several pathways—all much faster than those for nitrogen. As a result, phosphorus is a less limiting nutrient in the photic zone, even though its concentration is roughly one-seventh that of nitrogen. As with nitrogen, some phosphorus may be deposited in ocean sediments or in coastal regions in the form of bird guano; these materials are eventually recycled through weathering and erosion back into the ocean.

Nutrient Distribution in the Ocean

The availability of nitrogen and phosphorus often limits primary productivity and the resulting photic zone biomass. As shown in Table 18-1, the open ocean constitutes roughly 90 percent of the total oceanic environment and is often considered a "biological desert" because of the paucity of dissolved nutrients within the photic zone. Typical vertical concentrations of dissolved nitrate and phosphate for different oceans are shown in Figures 18-9. These nutrient patterns are the direct result of the biogeochemical cycling discussed above. Dissolved nutrient concentrations are low in the photic zone because available nitrogen and phosphate are quickly incorporated into the living

biomass through primary productivity. In addition, significant amounts of nutrients are exported from the photic zone through the continuous organic "rain" of fecal material, carcasses, and molts. while bacterial decomposition of this organic "rain" continues beneath the photic zone, the released nutrients are effectively sequestered from the photic zone by the strong density difference between the mixed layer and the deep-water layer, as marked by the thermocline. Thus, if vigorous productivity is to take place, these nutrient-rich deep waters must somehow be transported upward to "fertilize" the nutrient-poor photic zone. Physical mechanisms that accomplish this fertilization include upwelling and the weakening of the thermocline during the winter season or intense storms. Note that the absolute concentrations of dissolved nitrogen and phosphorus are higher in the deep waters of the Pacific and Indian oceans than in the Atlantic. This is because bottom waters in the Pacific and Indian oceans have been in the thermohaline circulation system for a longer time and have thereby accumulated more dissolved nutrients from overlying surface waters.

Oceanographers often construct "box models" to summarize how nutrients are cycled through different parts of the ocean through photosynthesis, respiration, mixing, runoff, and burial

TABLE	18-1

Productivity and fish production of the ocean

Area	Percentage of ocean	Area (square kilometers)	Average productivity (grams of carbon per square meter per year)	Average number of trophic levels (approximate)	Annual fish production (metric tons)
Open ocean	90	326,000,000	50	5	160,000
Boundary-current and open-ocean upwelling areas*	9.9	36,000,000	100	3	120,000,000
Coastal upwelling areas	0.1	360,000	300	1.5	120,000,000
Total annual fish production					240,160,000
Amount available for sustained harvesting†					100,000,000

Source: After Ryther, Science, 1969.

*Including certain offshore areas where hydrographic features bring nutrients to the surface.

†Not all the fish can be taken; many must be left to reproduce or the fishery will be overexploited. Other predators, such as seabirds, also compete with us for the yield.

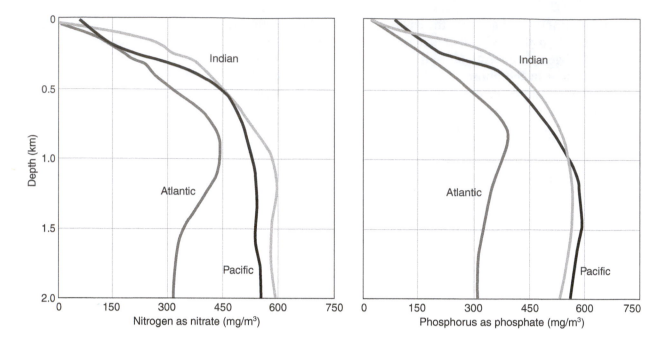

Figure 18-9 Curves showing the vertical distribution of dissolved nitrate and phosphate in seawater from non-upwelling oceanic regions of the Atlantic, Pacific, and Indian Oceans. [After Gifford B. Pinchot, "Marine Farming." Copyright © 1970 by Scientific American, Inc. All rights reserved.]

processes. An example of a box model for phosphorus is given in Figure 18-10, where the ocean is divided into two boxes (photic zone and deep sea) with exchanges within and between these boxes shown as arrows. Phosphorus enters the ocean from river runoff and exits from the ocean through burial in sediments. The numerical values in parentheses show the proportion of phosphorus cycled by each process. As the figure shows, for each atom added to the photic zone box by river runoff, approximately 99 atoms are upwelled. These 100 atoms are rapidly recycled in the photic zone through photosynthesis and respiration (including decomposition), but photosynthesis slightly exceeds respiration, and over approximately ten cycles a total of 100 atoms will sink out of the photic zone in the form of dead organic matter or waste. Ninety-nine of these 100 atoms will be respired, but one atom will be buried in sediments. This system is said to be in *steady state* because for each atom lost from the system by burial, another replaces it through river runoff. Note also that primary productivity would eventually cease if phosphorus removal from the photic zone was not balanced by upwelling of nutrient-rich wa-

ters. Similarly, respiration in the deep sea would consume all available oxygen if oxygen consumption was not balanced by the sinking of cold, oxygen-rich surface waters at high latitudes (see Exercise 8).

Nutrient Supply and Productivity

Only about 10 percent of the ocean has a reasonable amount of primary productivity and significant fish production (Table 18-1). In fact, only 0.1 percent of the ocean's environment produces about 50 percent of the fish available for human harvest. This rather startling statistic is a function of the average nutrient supply, primary productivity, and number of trophic levels in each environment. As demonstrated in the table, the open ocean environment is about 15 percent as productive as the boundary-current and open-ocean upwelling environments, and these are about 65 percent as productive as coastal upwelling environments. However, the coastal upwelling environment produces as many metric tons of fish as do the boundary-current and open-ocean upwelling environ-

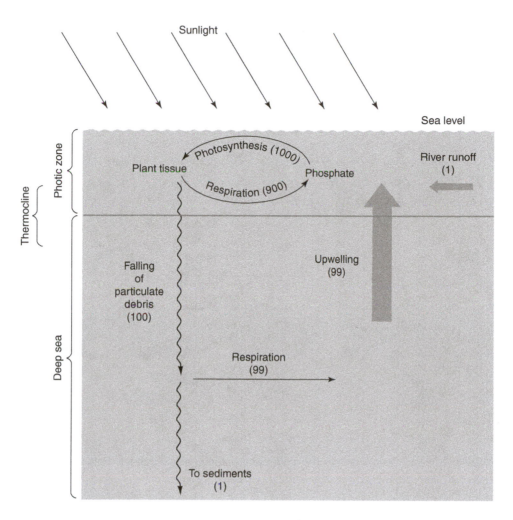

Figure 18-10 Box model of phosphorus cycling in a two-box ocean by photosynthesis, respiration, mixing, runoff, and burial processes. Numbers in parentheses represent phosphorus input, output, and exchange by different processes.

ments, while the upwelling environments as a whole produce 1500 times as much fish as does the remaining 90 percent of the ocean. The main reasons for these differences are the supply of "fresh" nutrients and number of trophic levels in each environment. Coastal upwelling regions have abundant nutrients supplied from below and an average of 1.5 heterotroph levels—their food webs are very short and simple (Figure 18-11a). In these areas, the constituents of the first trophic level are usually aggregates of colonial diatoms that are large enough to be fed upon directly by harvestable fish. In boundary-current or open-ocean upwelling regions, nutrients are more limited, and there are a greater number of heterotroph levels, with solitary diatoms fed upon by copepods, and copepods fed upon by harvestable fish (Figure 18-11b). Finally, the open-ocean region has very low nutrient replenishment and an even longer trophic web: solitary diatoms are eaten by microplankton (e.g., radiolarians), microplankton are eaten by mesoplankton (e.g., copepods, chaetognaths), mesoplankton are eaten by small, nonharvestable fish, and these small fish are eaten by harvestable fish, such as tuna (Figure 18-11c).

Why does nutrient supply and the number of trophic levels influence ultimate fish production? Lower nutrient supply limits primary productivity, which limits total productivity. Also recall the phenomenon of ecological efficiency discussed above—only about 10 percent of the energy contained in a trophic level is transferred up to the

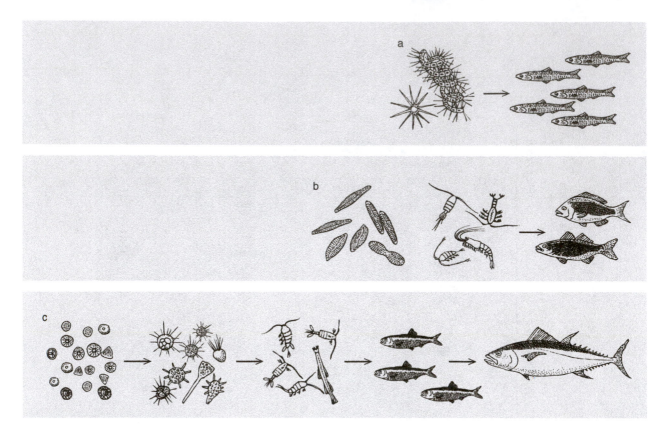

Figure 18-11 Comparison of the length and makeup of food chains from the following areas: (a) high-productivity coastal waters; (b) a boundary-current upwelling area; (c) low-productivity open-ocean waters. [After Gifford B. Pinchot, "Marine Farming." Copyright © 1970 by Scientific American, Inc. All rights reserved.]

next trophic level because about 90 percent is used for respiration, growth, and reproduction. Therefore, each intervening trophic level between primary producers and harvestable fish incurs a *10-fold decrease* in the biomass of harvestable fish compared to the biomass of primary producers. Thus, the longer the trophic web, the less "efficient" the ecosystem is from the standpoint of harvestable fish.

DEFINITIONS

Autotrophs. Organisms that produce their own food from inorganic matter.

Decomposers. Organisms that obtain their food from dead organic matter. Decomposers are a particular type of heterotrophs.

Ecological efficiency. Efficiency with which energy is transferred from one trophic level to the next higher level. Usually expressed as a ratio or percentage, it is the amount of living matter added to a trophic level compared to the amount of living matter required to produce it.

Heterotrophs. Organisms that obtain their food from other organisms.

Nutrients. Elements, such as phosphorous and nitrogen, necessary for life processes.

Trophic pyramid or **web.** A summary of the ways in which organisms within an ecosystem obtain their energy, either from inorganic matter (autotrophs) or from other living creatures (heterotrophs).

Exercise 18

Marine Ecosystems and Nutrient Cycles

NAME _____

DATE _____

INSTRUCTOR _____

1. (a) Propose two physical oceanographic phenomena that could cause a poor "crop" of diatoms during a given year in the Antarctic Ocean. _____
_____
_____
_____

(b) How would this affect the biomass at higher trophic levels? _____
_____
_____
_____

2. Trophic webs can be very complex and interconnected. From Figure 18-3, determine the possible population response of the following organisms to an overpopulation of skuas. Provide reasoning for your responses.

(a) Adélie penguin (one possible response): _____
_____

(b) crabeater seal (two possible responses): _____
_____

(c) leopard seal (two possible responses): _____
_____
_____

3. Assume that the cormorant population in the Long Island estuary ecosystem (Figure 18-4) was decimated by a cormorant-specific virus. For each group of animals below, state what would initially happen to their populations. Provide reasoning for your responses.

(a) flukes and eels: _____
_____

(b) water plants: _____
_____

(c) osprey and mergansers: _____
_____
_____

4. What would you hypothesize would eventually happen to the Long Island estuary ecosystem after the cormorant-specific virus passed? _____
_____
_____
_____
_____

5. How might you modify the Long Island estuary ecosystem to produce the following population effects?
   (a) decrease blowfish abundance; maintain fluke abundance _____
   _____
   _____

   (b) decrease tern abundance; maintain osprey abundance _____
   _____
   _____

6. What might happen to a given ecosystem if the top carnivore biomass decreased due to disease? _____
_____
_____

Why? _____
_____
_____
_____
_____

7. Why is a more complex ecosystem (i.e., greater number of pathways) probably more stable than a simple ecosystem? _____
_____
_____
_____
_____

8. Answer the following questions on ecological efficiency using the food web outlined for the Long Island estuary (Figure 18-4). For our purposes, assume that the water plants and marsh plants constitute the first trophic level and that the organic detritus and plankton (copepods and diatoms) constitute the second trophic level.
   (a) Rank each bird according to its ecological efficiency. Thus, the first bird listed will obtain its food by the shortest trophic pathway (fewest levels) from the first trophic level. To calculate how directly a given species is dependent upon a particular prey item, assume that each thick line represents one biomass unit and each thin line represents one-half biomass unit. For example, the tern gets one unit from the silversides and a half unit from the billfish. Some birds will share the same rating in ecological efficiency.

   1._____   5._____
   2._____   6._____
   3._____   7._____
   4._____   8._____

   (b) Assuming all other factors (i.e., reproduction rate, predation, etc.) equal, which bird species would you predict to be the most abundant and which the least abundant in the Long Island estuary area?
   _____

   Why? _____
   _____
   _____
   _____

9. The original purpose of the Long Island estuary study was focused on documenting DDT concentration. The numbers beside each species indicate the average parts per million (ppm) of DDT found in each species.

(a) What is the average amount of biological magnification of DDT from the first trophic level to the second trophic level? (As in question 8, assume that the water plants and marsh plants constitute the first trophic level.) _____

_____

_____

(b) From the observed DDT concentrations, hypothesize which prey item the blowfish consume the most. Explain your reasoning. _____

_____

_____

(c) Terns and mergansers are members of the fourth trophic level, but contain quite different DDT concentrations. Hypothesize two possible reasons for this difference, and state how you might test your proposed reasons. (Hint: The average age of a tern is about half that of a merganser.) _____

_____

_____

_____

(d) Compare your ranking of the ecological efficiency of the various birds (Question 8a) and the average concentration of DDT in their tissues. Does there appear to be any correlation? _____ If so, explain the process that would produce such a pattern. _____

_____

_____

_____

_____

10. In the Aleutian Islands of the North Pacific, Dr. J. A. Estes and co-workers at the University of California at Santa Cruz documented an abrupt decline in sea otters, which they attributed to increased predation by killer whales. While killer whales have always been a top carnivore in the open-ocean ecosystem, they appear to have recently shifted their trophic strategy to prey upon sea otters living within the shallow, nearshore ecosystem. Data from the study are provided on the following page. The basic food chain of sea otter-sea urchin-kelp is presented on the left, and the recent addition of the killer whale to the trophic chain is shown on the right. The relative sizes of the arrows indicate the relative biomass consumption before and after the addition of killer whales to the ecosystem. The four graphs show available data on changes in sea otter abundance, sea urchin biomass, urchin grazing intensity, and total kelp density from 1972 to 1997. Use these data to address the following questions:

(a) Describe each organism in the food chain using the trophic level terms of first-level carnivore, top carnivore, primary producer, and first-level herbivore. _____

_____

_____

_____

(b) Hypothesize some possible events that could have caused killer whales to shift their trophic strategy toward sea otters, which previously were not a major food item. What data would you need to test your hypotheses? _____

_____

_____

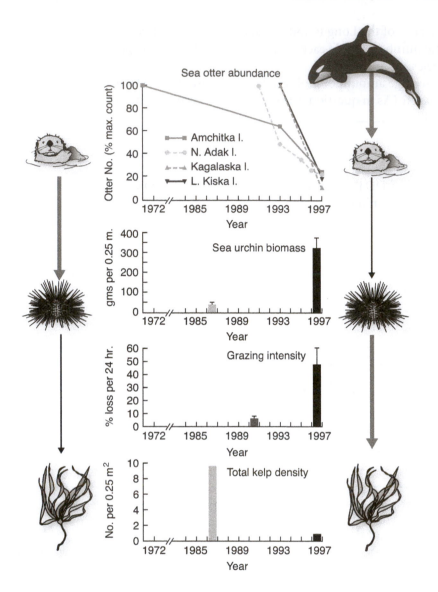

(c) Describe the effects that the addition of killer whales to the nearshore food chain have had on the relative biomass of organisms in each of the lower trophic levels. _____
_____
_____

(d) In addition to the organisms shown, many others are dependent upon kelp forests for shelter, reproduction, food, etc. What may happen to this ecosystem if the observed trend continues? _____
_____
_____

(e) Estes et al. discuss their data and their implications with the caveat that their data are not as complete as they would like, although the general trends are clear. What additional data would you collect in the future from the nearshore or oceanic ecosystem to better understand this trophic web change?_____
_____

11. The structure of, and interactions within, a given ecosystem can change over time scales ranging from thousands of years to hourly. Changes on hourly scales commonly occur in the intertidal zone, where the environment fundamentally changes as tides rise and fall. The general structure of salt marsh ecosystems on the Atlantic coast during high and low tide is shown below. Examine how the ecosystem structure changes through a tidal cycle and answer questions on the following page.

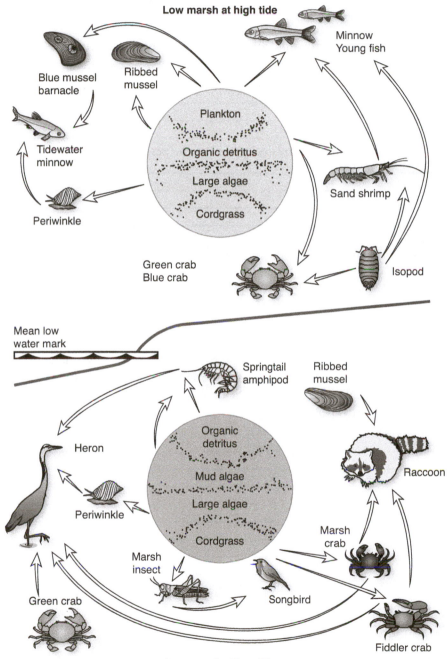

(a) Which organisms are involved in the food web during both high and low tides? _____
_____
_____
_____

(b) Which organisms are the top carnivores during high tide? During low tide? _____
_____
_____
_____

(c) The loss of which organisms would affect the structure of the salt marsh ecosystem during both high
and low tides? _____
_____
_____
_____

(d) Which would be more affected by decimation of the periwinkle population, the high or low tide
ecosystem? _____
_____
_____
_____

(e) Why does the ribbed mussel not show any feeding activity during low tide? _____
_____
_____
_____

12. In general, the rate of respiration must nearly balance that of photosynthesis in the ocean as a whole.
However, over geologic time, there has been a slight excess of photosynthesis over respiration. Discuss how
this fact would affect the following over geologic time:

(a) composition of the atmosphere: _____
_____
_____
_____

(b) formation of natural petroleum resources: _____
_____
_____
_____

(c) input rate of nutrients to the ocean versus their burial rate: _____
_____
_____
_____

Plankton

OBJECTIVES:

■ To discover the importance of plankton in marine ecosystems.

■ To classify and identify some important animal and plant plankton.

Plankton are free-floating pelagic organisms that are unable to swim against a current of about 1 knot for an extended period of time. Most plankton are relatively small in contrast to the larger, stronger **nekton,** such as squid, fish, and whales, which are effective swimmers. Plankton are extremely important in the marine ecosystem, as they include most of the primary producers and herbivores that form the base of most food pyramids (see Exercise 18).

Classifying Plankton

Plankton may be classified in a number of ways. They may be divided by trophic level into **phytoplankton** (plant plankton) or **zooplankton** (animal plankton). They may also be classified by habitat; for example, by their relative distance from land, such as neritic (nearshore) versus oceanic (offshore), or by their relative depth in the water column, such as epipelagic (surface to 100–200 meters) versus mesopelagic (200–1000 meters). In addition, special terms are used for plankton occupying specific habitats; for example, *neuston* refers to plankton that live at the ocean-atmosphere boundary. From a life-cycle standpoint, the term **holoplankton** applies to organisms that remain plankton throughout their lives, whereas **meroplankton** are plankton only during the early larval to juvenile stages of their lives. Finally, plankton can be classified according to size, as in the following list:

Category	Size range	Example size
Megaplankton	>2000 micrometers	Larger than a nickel's diameter
Mesoplankton	200–2000 micrometers	Smaller than a nickel's diameter
Microplankton	20–200 micrometers	Smaller than the letter "o"
Nanoplankton	2–20 micrometers	Roughly pin-head size
Ultrananoplankton	<2 micrometers	Smaller than a pin head

In this exercise you will examine a variety of plankton, try to identify them using the above classifications, and hypothesize a general planktonic food chain based on your observations and reading. First, you will need to acquire some plankton! If you live near the ocean, you and/or your instructor may be able to collect living plankton. If not, preserved plankton may be purchased from biological supply houses—try to get a variety of samples from different environments and regions. If you live near a lake or pond, you may want to collect freshwater plankton to observe their activities and interactions. If you do so, you will notice that freshwater plankton tend to be less diverse than their marine counterparts.

Observation of Living Plankton

Observing living plankton can be a fascinating experience, particularly if you have never seen microscopic organisms. Gently shake the plankton bottle to mix the plankton, and pour a small volume of water into a petri dish. Place the dish on the stage of a dissecting microscope with light coming from below, and observe the plankton at low magnification ($\sim 20\times$). After focusing up and down, you will see that most plankton are at the bottom of the dish, while some are concentrated at the surface. An illustrated guide to common marine plankton is provided in Figure 19-1. Study the organisms using different magnifications, but appreciate that they may be hard to track at higher magnifications. You may try to slow them down by adding cotton fibers, gelatin, or Protoslo to the sample. You instructor may have special dye available to make visible the currents produced by the plankton's locomotion and feeding; the latter is particularly noticeable in crustaceans that generate currents to bring food to their mandibles. Observe and note any trophic interactions among the plankton (i.e., who is eating whom); such information will help you in constructing your plankton ecosystem diagram.

Notice that some of the phytoplankton move—something you wouldn't normally expect from more familiar photosynthesizers such as your houseplants! Most phytoplankton, such as the dinoflagellates and flagellates, use their flagella—whip-like extensions—to move about, whereas pennate diatoms glide along by creating waves in their cell membrane along the groove between their valves. Zooplankton may glide or dart about through the use of cilia (hair-like extensions from the cell surface), body undulations, or swimming appendages. Many zooplankton moving about by the latter two modes are meroplanktic forms of benthic and nektonic organisms. Some of the larger zooplankton you may see include mysids, shrimps, chaetognaths (glass or arrow worms), coelenterate medusae, and copepods. The copepods are among the most common of zooplankton and use their antennae to swim; in fact, the word *copepod* is derived from Greek, meaning "oar-foot."

Attempt to observe some micro- and nanoplankton using a higher-powered light transmission microscope. Remove some water from the plankton bottle's bottom with an eyedropper and place a few drops on a depression slide or glass slide with a cover slip. If smaller plankton are present, they will appear to be moving rapidly due to the small field of view; you may want to add a small amount of gelatin or Protoslo to slow them to an observable rate.

Observation and Identification of Preserved Plankton

When working with preserved plankton samples, be careful to avoid contact with the liquid. It may contain formalin, a preservative to which some people are allergic. You may prefer to view the plankton in a petri dish with a glass cover to avoid the fumes. Regardless, be sure that the lab room is well ventilated. If the glass cover fogs up, rub some mineral oil on the bottom. If the smell is intolerable, your instructor should have some fine sieves to wash the samples off with fresh water. To observe nanoplankton, use an eyedropper to collect a few drops from the bottom of a preserved plankton bottle that has been sitting for a few hours. Observe the sample, looking for the various characteristics described in the living plankton discussion and Figure 19-1.

Density Calculation

To determine the concentration or **standing crop** of plankton, shake the original sample vigorously, remove a known volume (10–50 milliliters of liquid), and allow it to stand for an hour. Carefully decant off the liquid except for a few milliliters, pour the remaining volume into a small petri dish, and then count the plankters. Divide this count by the percentage of the original sample volume you examined. For example, if you sampled 10 milliliters from a 1000-milliliter volume and found 112 plantkers, dividing 112 by 1 percent (10 milliliters/1,000 milliliters) provides an estimate of the plankton concentration in the sample: about 11,200/liter.

The next step is to determine the volume of seawater that was filtered for the original sample. Calculate this seawater volume by multiplying the surface area of the mouth of the collecting net by the total distance over which it was towed. For

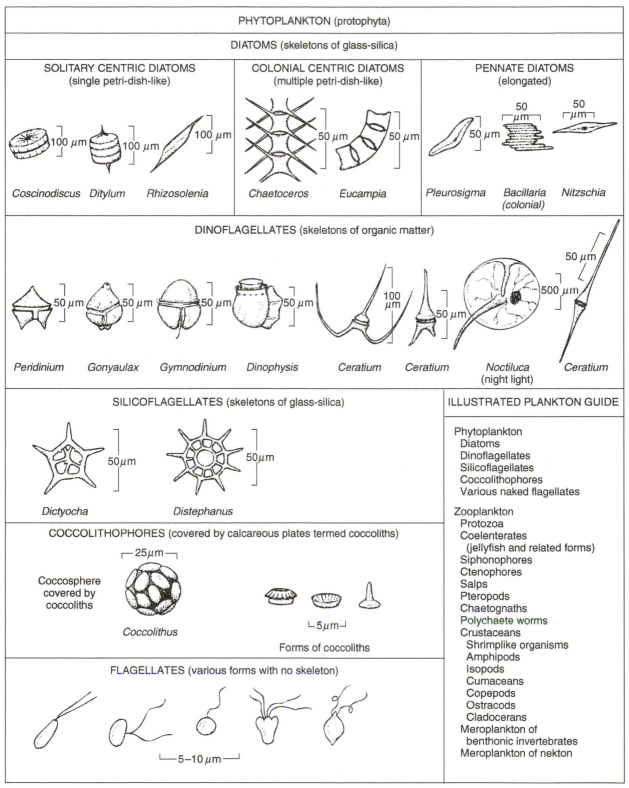

(continued)

Figure **19-1** A pictorial guide to common types of marine plankton.

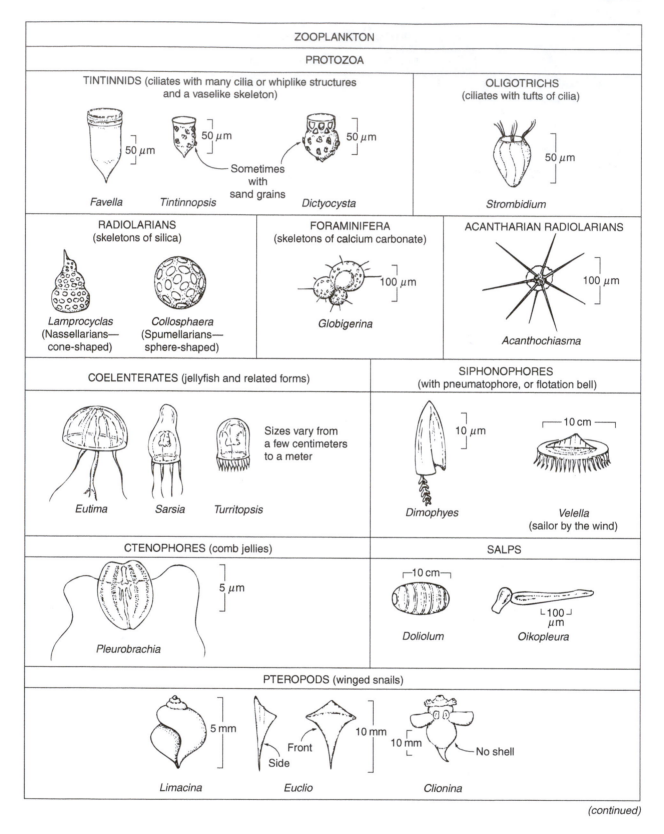

ZOOPLANKTON

PROTOZOA

TINTINNIDS (ciliates with many cilia or whiplike structures and a vaselike skeleton)

50 μm

50 μm

50 μm

Sometimes with sand grains

Favella   Tintinnopsis   Dictyocysta

OLIGOTRICHS (ciliates with tufts of cilia)

50 μm

Strombidium

RADIOLARIANS (skeletons of silica)

Lamprocyclas (Nassellarians— cone-shaped)

Collosphaera (Spumellarians— sphere-shaped)

FORAMINIFERA (skeletons of calcium carbonate)

100 μm

Globigerina

ACANTHARIAN RADIOLARIANS

100 μm

Acanthochiasma

COELENTERATES (jellyfish and related forms)

Sizes vary from a few centimeters to a meter

Eutima   Sarsia   Turritopsis

SIPHONOPHORES (with pneumatophore, or flotation bell)

10 μm

10 cm

Dimophyes   Velella (sailor by the wind)

CTENOPHORES (comb jellies)

5 μm

Pleurobrachia

SALPS

10 cm

100 μm

Doliolum   Oikopleura

PTEROPODS (winged snails)

5 mm

Front

Side

10 mm

10 mm

No shell

Limacina   Euclio   Clionina

(continued)

CHAETOGNATHS (arrow or glass worms)

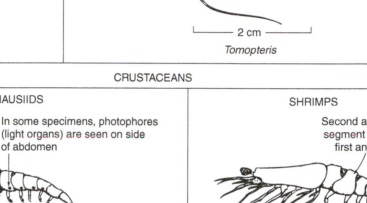

— 1 cm —

Sagitta
(tiger of the sea)

POLYCHAETE WORMS

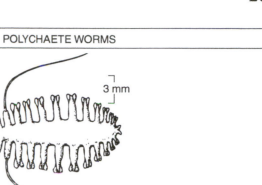

3 mm

— 2 cm —

Tomopteris

CRUSTACEANS

EUPHAUSIIDS

In some specimens, photophores (light organs) are seen on side of abdomen

Exposed gills

Abdominal legs

— 1–6 cm —

Euphausia

SHRIMPS

Second abdominal segment overlaps first and third

Legs on abdominal segments

— 1–10 cm —

Pasiphaea

MYSIDS

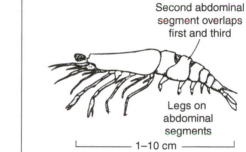

Shell (the carapace – hard outer covering) is not attached to the abdomen as in shrimp and euphausiids; most mysids are smaller (except in deep sea) than shrimp and euphausiids

Reduced legs on abdomen

Statocyst (round) equilibrium organ) on tail

— Less than 1–2 cm —

Praunus

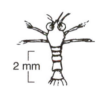

2 mm

The juvenile or "mysid" stage; many crustaceans have a similar appearance at this stage of development

AMPHIPODS (Unstalked eyes and laterally compressed; no carapace)

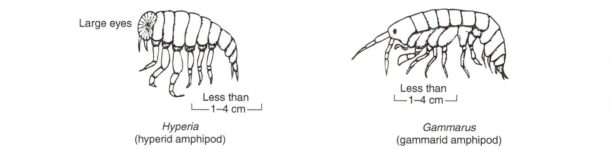

Large eyes

Less than
1–4 cm

Hyperia
(hyperid amphipod)

Less than
1–4 cm

Gammarus
(gammarid amphipod)

(continued)

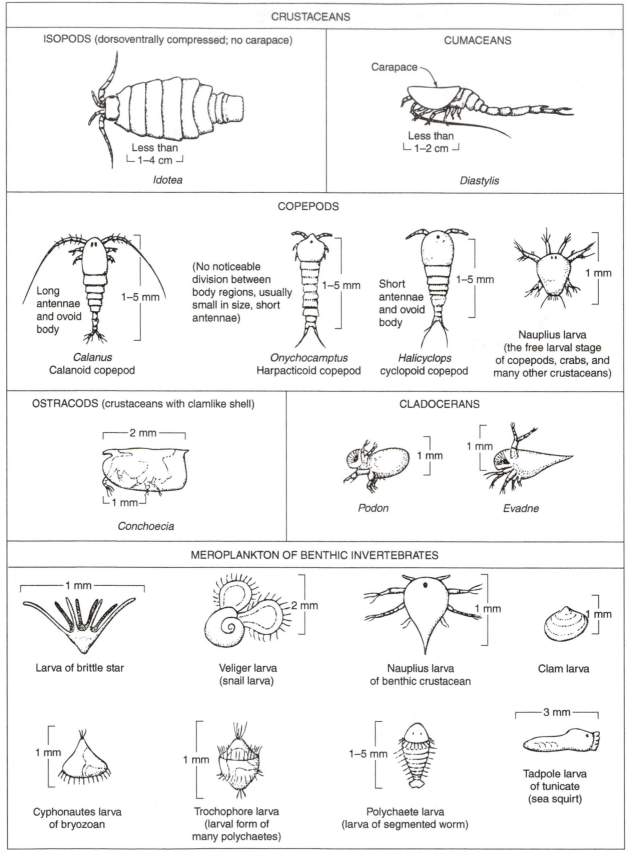

CRUSTACEANS

ISOPODS (dorsoventrally compressed; no carapace)

Less than
1–4 cm

Idotea

CUMACEANS

Carapace

Less than
1–2 cm

Diastylis

COPEPODS

Long antennae and ovoid body

1–5 mm

Calanus
Calanoid copepod

(No noticeable division between body regions, usually small in size, short antennae)

1–5 mm

Onychocamptus
Harpacticoid copepod

Short antennae and ovoid body

1–5 mm

Halicyclops
cyclopoid copepod

1 mm

Nauplius larva
(the free larval stage of copepods, crabs, and many other crustaceans)

OSTRACODS (crustaceans with clamlike shell)

2 mm

1 mm

Conchoecia

CLADOCERANS

1 mm

Podon

1 mm

Evadne

MEROPLANKTON OF BENTHIC INVERTEBRATES

1 mm

Larva of brittle star

2 mm

Veliger larva
(snail larva)

1 mm

Nauplius larva
of benthic crustacean

1 mm

Clam larva

1 mm

Cyphonautes larva
of bryozoan

1 mm

Trochophore larva
(larval form of
many polychaetes)

1–5 mm

Polychaete larva
(larva of segmented worm)

3 mm

Tadpole larva
of tunicate
(sea squirt)

(continued)

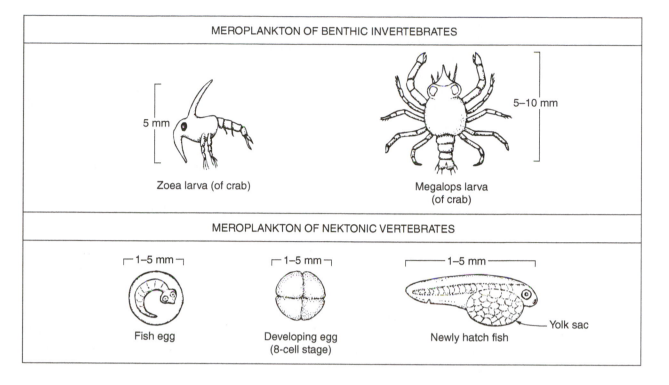

MEROPLANKTON OF BENTHIC INVERTEBRATES

5 mm

Zoea larva (of crab)

5–10 mm

Megalops larva
(of crab)

MEROPLANKTON OF NEKTONIC VERTEBRATES

1–5 mm

Fish egg

1–5 mm

Developing egg
(8-cell stage)

1–5 mm

Yolk sac

Newly hatch fish

simplicity, assume that filtering was 100 percent efficient. This assumption may not always be true, particularly if the plankton net was towed too quickly and a turbulent "head" of water formed at the net's mouth. Dividing the estimated total number of plankters in the original sample by the total volume of water sampled will provide an estimate of the standing crop of plankton. If you are interested in determining the standing crop of each different type of plankton, you may need to use smaller subsamples to make the process manageable in the given amount of lab time.

Counting the Plankton

On the counting sheet provided in Question 1, count the different plankters from your preserved sample, using Figure 19-1 or, even better, a detailed field key or other literature specifically for the sample region, to identify them. Identify the individuals to the species level if possible. If you cannot assign a species name to a distinct plankter, assign it a temporary name (e.g., centric diatom #1) and provide a detailed sketch, which will allow you to compare your results with fellow students' sam-

ples. At the top of the counting sheet, there is space for information on where, when, and how the sample was collected; for example, you should specify the net type, mesh size, water volume filtered, and the exact locale and depth. If you are using material from a biological supply house, however, some of this information may not be available.

If you or the class are examining multiple plankton tows, note the results of each on a separate sheet, compare them, and see if you can explain any differences. For example, a nearshore sample may contain more meroplankton than an offshore sample, or a sample from near a sewage outfall may have lower diversity, but greater total abundance, than a sample from farther away.

Diagramming a Plankton Trophic Web

Review the principles of a trophic web in Exercise 18 or your course text. In Question 2, you will construct a planktonic trophic web by placing species in their hypothesized trophic levels in the plankton ecosystem. For example, the base of the web would consist of the primary producers or phytoplankton, so you would write down all of your identified phytoplankton along the base line. For the next level, primary consumers, write down those zooplankton that feed directly on phytoplankton, such as copepods. Construct as many trophic levels as you hypothesize to exist in your sample.

The more you observe the living plankton sample, the more accurate your trophic web will become. However, your sample is probably not a *closed* ecosystem—one that is self-sufficient and reasonably uninfluenced by other organisms. Rather, you probably have an *open* ecosystem, in which the actual top carnivores are fish or larger animals not sampled by the plankton tow. Conversely, you may be missing smaller plankton if a relatively coarse net mesh was used. After placing each plankter in its hypothesized trophic level, add in the other organisms you would expect to be present if you had a complete sampling of organisms living in the water column (i.e., a closed ecosystem). Appreciate that an ecosystem requires a source of energy and inorganic nutrients; incorporate these into your trophic web as well.

DEFINITIONS

Holoplankton. Organisms that remain as plankton throughout their lives.

Meroplankton. Organisms that are plankton only during the early larval to juvenile stages of their lives.

Nekton. Organisms that actively swim within the water column, generally by virtue of their relatively large size.

Phytoplankton. Free-floating organisms that produce their own food through photosynthesis.

Plankton. Organisms, generally relatively small, that live in the water column and are free-floating, unable to swim against a mild current (~ 1 knot).

Standing crop. The abundance of a particular type of organism or organisms at a given moment, generally calculated as a concentration (e.g., number per liter, mass per liter).

Zooplankton. Free-floating organisms that depend upon other organisms for their energy needs.

Exercise *19*

*P*lankton

NAME _____

DATE _____

INSTRUCTOR _____

1. On the plankton counting sheet provided on the following page, list each species name under the appropriate category, if possible, and/or draw and count each plankter type in your sample. Try to work at the species level; even if you can't name the species, you can designate distinct forms by numbers (e.g., centric diatom #1, centric diatom #2).

2. (a) Hypothesize the structure of the plankton food web by placing the name or drawing of each species in your sample into the appropriate trophic level. Include as many trophic levels as you think exist in your sample. Six trophic levels are provided; however, you may not be able to fill all six or may have to add more depending on the diversity of your plankton sample.

Top carnivores (Trophic level 6)

Third-level carnivores (Trophic level 5)

Second-level carnivores (Trophic level 4)

First-level carnivores (Trophic level 3)

PLANKTON COUNTING SHEET
STATION_____ LOCATION (locale and depth)_____
EQUIPMENT USED _____ DATE AND TIME OF COLLECTION_____
AMOUNT OF TOTAL SAMPLE COUNTED (in density calculation)_____

PHYTOPLANKTON

Diatoms	Dinoflagellates	Other phytoplankton
_____	_____	_____
_____	_____	_____
_____	_____	_____
_____	_____	_____
_____	_____	_____
_____	_____	_____

ZOOPLANKTON

Protozoa	Coelenterates	Crustaceans (adults)
_____	_____	_____
_____	_____	_____
_____	_____	_____
_____	_____	_____
_____	_____	_____
_____	_____	_____

Crustaceans (larvae)	Other holoplankton	Meroplankton
_____	_____	_____
_____	_____	_____
_____	_____	_____
_____	_____	_____
_____	_____	_____

Herbivores (Trophic level 2)

Primary producers (Trophic level 1)

(b) Sketch in the trophic or feeding pathways from one species to another, showing the various predator–prey relationships. If you are certain of a particularly pathway—for example, if you saw a copepod eat a centric diatom or a copepod inside a coelenterate—connect the pair with a *solid* arrow with its head pointing in the direction of energy flow (i.e., from prey to predator). If you are unsure of a particular pathway, but have good reason to think that it exists, draw a *wiggly* arrow to indicate high probability, and a *dashed* arrow to indicate that the relationship is possible.

(c) Assuming that your plankton sample captures only a portion of the total ecosystem, add the organisms you think would be present in the natural environment. Use a pencil of a different color.

(d) Complete your trophic web by indicating below trophic level 1 the inorganic ingredients necessary for primary production. Also indicate, using arrows and text on the side of the diagram, how, within the various trophic levels, dead organisms and fecal matter are converted back into nutrients by decomposers, such as bacteria (see Exercise 18).

3. (a) After identifying, counting, and assigning trophic levels to the various plankters in your sample, you should have a general idea of the quantity of organisms at each trophic level. Assuming that your sample accurately represents the natural community, compare the standing crops at each level. For example, are there more primary producers than herbivores? more herbivores than primary carnivores? If you have quantitative data, calculate densities (e.g., ~ 200 primary producers per liter). Your instructor will tell you how to register this information.

(b) In most natural ecosystems, only about 10 percent of the energy contained in one trophic level is transferred to the next higher level. Are your data consistent with this general pattern? _____
If not, give some possible reasons. _____
_____
_____
_____

(c) Through natural selection, planktonic organisms have evolved numerous adaptations to their environment, such as appendages to aid in floatation and locomotion. List as many specific examples of planktonic adaptations as you can from your plankton observations. Be specific, describing the structures involved and how they function. _____
_____
_____
_____
_____
_____

Marine Adaptations and the Deep–Sea Environment

OBJECTIVES:

■ To discover the adaptations that help marine organisms to cope with stressful deep-sea environments.

■ To investigate the ways in which daily vertical migrations protect, and provide a food supply for, marine organisms.

The processes by which an organism apparently becomes better suited to its environment, or for particular functions within that environment, are termed adaptations. These adaptations may be structural, biochemical, or behavioral, and they develop in the course of geologic time in response to physical and biological stresses. Here we will be investigating mainly those structures or activities of marine organisms that tend to equip them better for life in the deep sea.

An **adaptation** is a modification of a previous structure, biochemical process, or behavioral pattern that occurs over the course of evolution. Commonly known examples of adaptations in terrestrial animals are hooves, which equip horses and cattle for roaming and feeding on grasslands, the prehensile tail enabling some monkeys to live in an arboreal habitat, and the long neck that allows the giraffe to feed on leaves from the scattered trees of the savanna region. However, most of us are less familiar with the adaptations of marine organisms, especially some of the most interesting—those of the deep sea.

Some organisms have traits that are used one way in one environment but are suitable for use in another way in another environment: these traits are termed **preadaptations,** because they are adaptations that have already equipped the organism to exist in a new environment. A marine example is the brittle star. Although brittle stars that live in the deep sea look much the same as their shallow-water relatives, they have been preadapted to the deep-sea environment. Their major preadaptations are their long appendages and their ability to feed on detritus, or disintegrated matter (including mud).

The Ocean Environments

For many species, both marine and terrestrial, it is possible to compare the ancestral stock with the descendants, observe what the adaptations are, and determine their function, if we know the environmental conditions in which ancestors and descendants lived. In this discussion, we will be concerned mainly with the adaptations of deep-sea organisms that exist in the different marine environments that prevail at different ocean depths. These organisms (or their ancestors) have evolved from shallow-water marine organisms. By comparing them with their shallow-water relatives, we can determine what these adaptations are and what they are used for in the deep-sea environment.

The characteristics of the various ocean environments are given in Table 20-1, and this information will aid you in determining some of the reasons that different species evolved in the way they did in order to survive in a particular environment. First, the shallow-water environment is the epipelagic zone shown in Figure 20-1 and listed in the table (for a description of the marine life zones, see also Exercise 15). We will assume this shallow area to be the ancestral zone containing the

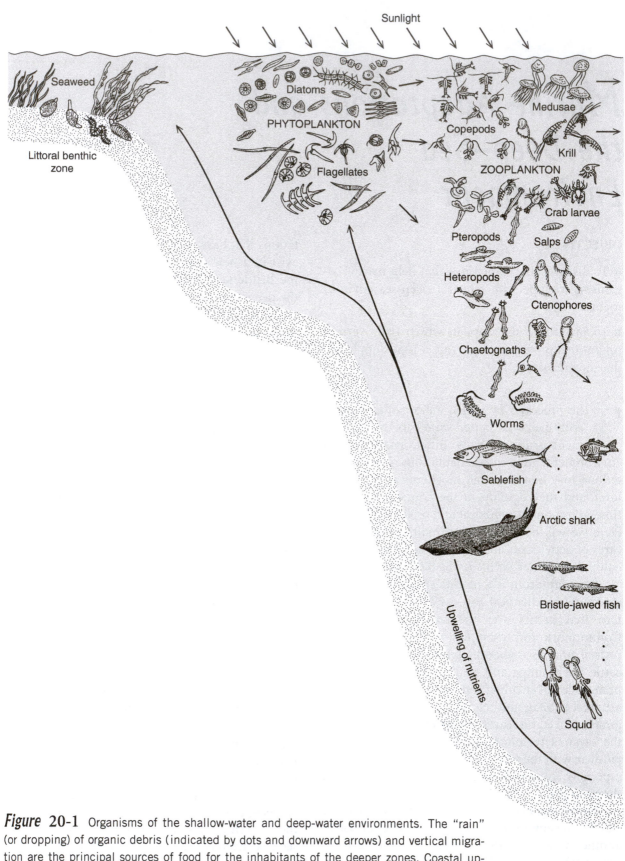

Figure 20-1 Organisms of the shallow-water and deep-water environments. The "rain" (or dropping) of organic debris (indicated by dots and downward arrows) and vertical migration are the principal sources of food for the inhabitants of the deeper zones. Coastal upwelling is indicated by the long arrows. Organisms are not drawn to the same scale. [After John D. Isaacs, "The Nature of Oceanic Life." Copyright © 1969 by Scientific American, Inc. All rights reserved.]

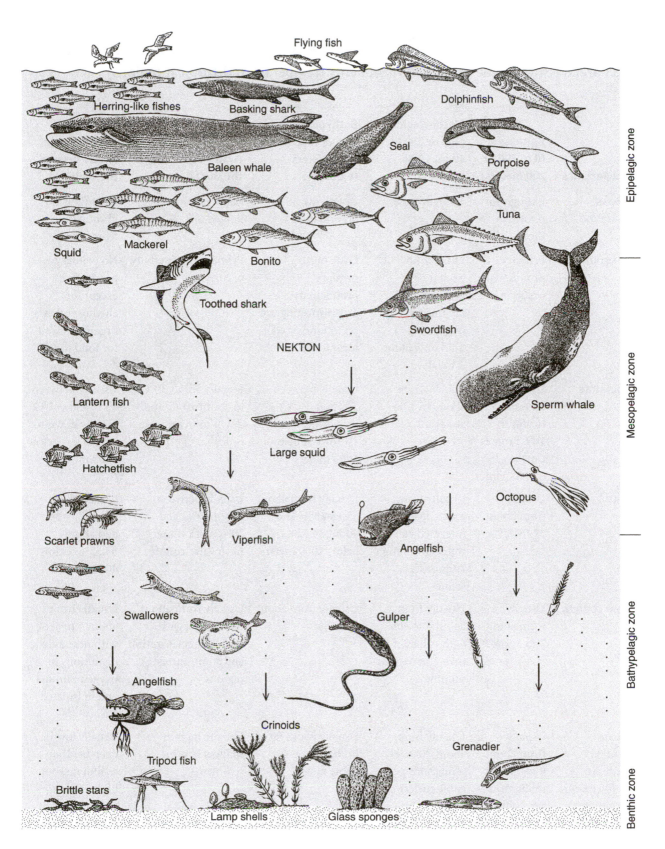

Flying fish

Herring-like fishes

Basking shark

Dolphinfish

Seal

Porpoise

Baleen whale

Tuna

Squid

Mackerel

Bonito

Toothed shark

Swordfish

NEKTON

Sperm whale

Lantern fish

Large squid

Hatchetfish

Octopus

Scarlet prawns

Viperfish

Angelfish

Swallowers

Gulper

Angelfish

Crinoids

Grenadier

Tripod fish

Brittle stars

Lamp shells

Glass sponges

Epipelagic zone

Mesopelagic zone

Bathypelagic zone

Benthic zone

TABLE   20-1

The characteristics of the marine environments (temperate and tropical regions)

Characteristics	Zones				
	Epipelagic (0 to 100 or 200 meters)	**Mesopelagic (100 or 200 to about 1000 meters)**	**Bathypelagic and deeper (about 1000 meters to bottom)**	**Shallow benthic (water over the shelf)**	**Deep benthic (water beyond the shelf)**
Degree of illumination	Enough for photosynthesis	Twilight zone	Essentially no illumination	Many portions lighted	Essentially no illumination from above
Food supply	Primary productivity occurring	Little or no primary productivity; organisms migrate up to food or wait for it to fall	Little or no primary productivity; organisms migrate up to food or wait for it to fall	Primary productivity occurring	No primary productivity except for chemosynthesis; organisms wait for food to fall
Temperature	Usually about 28°C to about 10°C; sometimes near 0°C in winter	Usually from about 15°C to about 5°C	Usually between 5°C and −2°C; usually down to 1°C or less below 4000 meters	Usually about 30°C to about 10°C, and down to freezing at times	Usually between 15°C and −2°C; usually down to 1°C or less below 4000 meters
Salinity	Usually from about 37 to 32‰	Usually from about 35−34.5‰; intermediate waters from high latitudes less saline	Usually from about 35−34.5‰ and about 34.52‰ below 4000 meters	Usually between about 40‰ and 30‰ with some freshwater runoff	Usually from about 35−34.5‰ and about 34.52‰ below 4000 meters
Oxygen content	Usually from about 7 to 3.5‰	Usually from about 5 to 4‰, with values of less than 1 in oxygen minima	Usually from about 6 to 5‰	Usually from about 7 to 3.5‰, with some supersaturation and some anoxic regions	Usually from about 6 to 4‰, with near anoxic conditions in oxygen minima and in isolated basins
Nutrient content (phosphate given for pelagic environments, and general organic carbon for the benthic environments)	Usually from about 0 to 30 milligrams per cubic meter; higher in upwelling regions	Usually from about 30 to 90 milligrams per cubic meter; higher in upwelling regions	Usually about 90 milligrams per cubic meter	Usually high in shallow benthic sediments	Usually low in deep benthic sediments, but high under upwelling regions

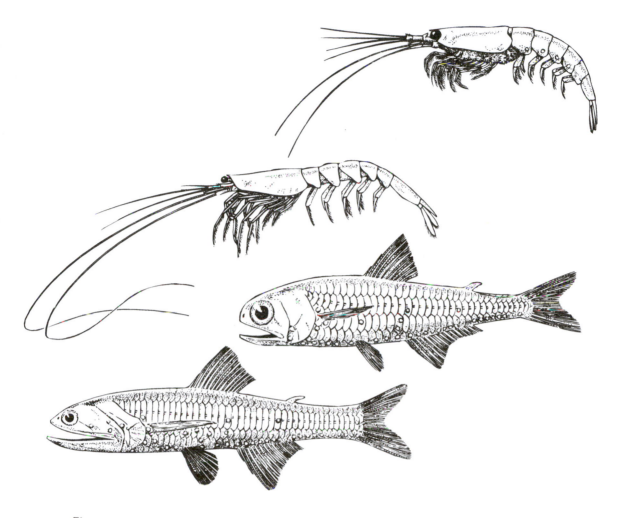

Figure 20-2 The organisms of the deep-scattering layer include crustaceans and fishes. Among them are (top to bottom) a euphausiid, a sergestid, and two forms of myctophid, or lantern fish. They range in size from an inch for the euphausiid to as long as 3 inches for the lantern fish. The four spots on the euphausiid and those on the myctophids are photophores. [After Robert S. Dietz, "The Sea's Deep Scattering Layers." Copyright © 1962 by Scientific American, Inc. All rights reserved.]

"standard" organisms from which the deep-water species evolved. But even in the epipelagic zone itself, we can find examples of adaptations from the "standard" fish type: one is the flying fish with its adaptive modification of fins to "wings" and its behavioral modification of "flying." Plankton, too, display adaptations to their environment.

Deep-Sea Adaptations

One important adaptation to the deep-sea environment is the ability to **bioluminesce** (to create and use biological light). Some organisms, such as squid, squirt an ink that luminesces as it comes in contact with the dissolved oxygen in the seawater. Other marine organisms have even more sophisticated mechanisms for bioluminescence, called **photophores,** or light organs. Some of these photophores are really nothing more than collections of luminescent bacteria, but others are highly evolved organs in which the light can be turned off and on. Many deep-sea fish use photophores for a variety of tasks. For example, the myctophid, or lantern fish, illustrated in Figure 20-2, swims in schools that migrate vertically through the water column, usually a few hundred meters every day. As the figure shows, these fish have large numbers of photophores. On this one species, the

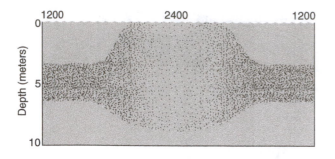

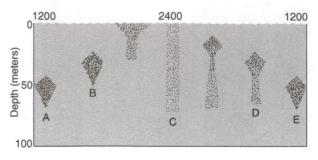

Figure 20-3 Phytoplankton migration over the course of a 24-hour period. Note that the plankton are spread out during the nighttime hours and packed together during the day. Hours of the day are noted in accord with the 24-hour maritime system.

Figure 20-4 Vertical migration pattern of the adult female copepod *Calanus finmarchicus.* The example demonstrates a pattern of migration whereby the population (A) clusters at midday, (B) migrates toward the surfaces at night, (C) disperses throughout the water column at midnight, (D) starts to reassemble at dawn, and (E) moves to its daytime depth at noon.

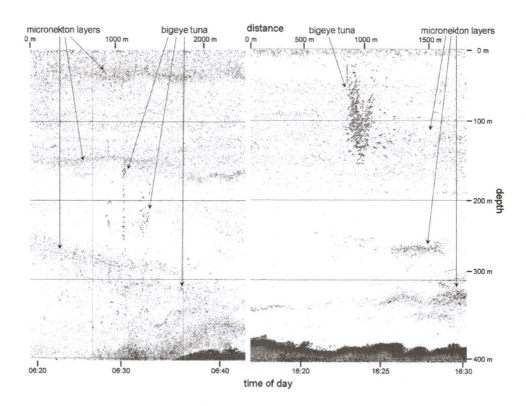

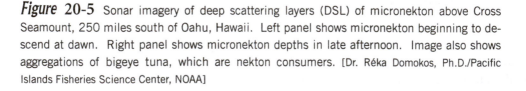

Figure 20-5 Sonar imagery of deep scattering layers (DSL) of micronekton above Cross Seamount, 250 miles south of Oahu, Hawaii. Left panel shows micronekton beginning to descend at dawn. Right panel shows micronekton depths in late afternoon. Image also shows aggregations of bigeye tuna, which are nekton consumers. [Dr. Réka Domokos, Ph.D./Pacific Islands Fisheries Science Center, NOAA]

photophores probably have a number of different functions.

Numerous other important structural adaptations can be seen in various deep-sea fish. One is the lack of scales on many species (scales assist the organism in swimming). Another is poorly developed musculature. Also, some deep-sea fish have developed a pattern whereby the male becomes totally dependent on the female for its existence, and many of its systems, except the reproductive system and a few others, degenerate.

Adaptations can also be biochemical or behavioral. A good example of a behavioral adaptation is that of **vertical migration.** Many kinds of marine organisms migrate vertically. Even some of the phytoplankton make short vertical migrations in the water column; an example is shown in Figure 20-3. You will notice that these phytoplankton cluster during the day (in response to the light) and disperse, or "wander," at night (when there is no light to which they can orient themselves).

One migration pattern that has been well studied is the vertical migration of the copepod *Calanus finmarchicus,* a herbivore and member of the second trophic level. The diatoms it eats live in the upper 50 meters of the water column. In turn, it is eaten by herring and similar species of fish that also occupy the upper 50 meters of water. Figure 20-4 illustrates the migration pattern of *Calanus finmarchicus* at one of its life stages. Notice that the population clusters at noon, migrates toward the surface at night, disperses throughout the water column at midnight, starts to reassemble at dawn except for some of the individuals at deeper levels, and returns to its so-called daytime depth at noon.

The phytoplankton migration and the *Calanus finmarchicus* migration are usually termed *shallow-water vertical migrations.* Another interesting migration is that of an entire community of various organisms from fairly deep water (about 500 meters) to the surface and back again. The density of these organisms is so high that they may reflect the sound waves sent out by bottom-finding

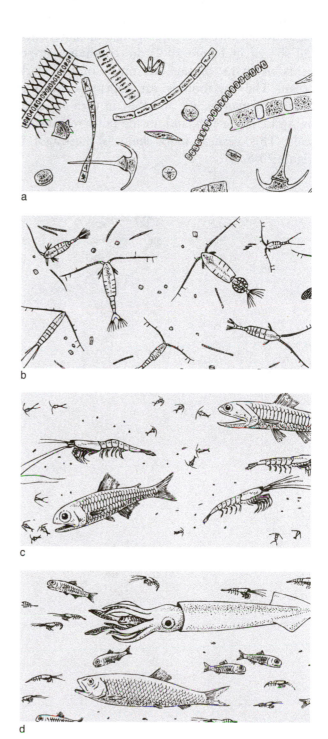

a

b

c

d

Figure 20-6 The deep scattering layer is an important link in the deep-sea food chain. (a) Phytoplankton are the "grass" of the sea. These are diatoms and dinoflagellates. (b) Copepods feed on phytoplankton and in turn furnish food for larger animals. (c) "Deep scatterers" rise from depths at night to feed in the epipelagic zone. The myctophids (lantern fish) eat copepods; the euphausiids consume the phytoplankton. (d) Larger animals, such as squid and herring, sometimes eat the organisms of the deep scattering layers at night near the surface. Fur seals are known to feed extensively on myctophids. [After Robert S. Dietz, "The Sea's Deep Scattering Layers." Copyright © 1962 by Scientific American, Inc. All rights reserved.]

oceanographic equipment, and be erroneously recorded on the receiver as the bottom (as they have been many times). Because these communities reflect the sound waves, occur in more than one layer, and frequently descend to deep levels, they have been given the name *deep scattering layers,* or DSL's. An actual recording of a migration of DSL's off the coast of Oahu is shown in Figure 20-5.

We do not know all the organisms that constitute the DSL's, but we do know that euphausiids, sergestid shrimp, and myctophids, or lantern fish (all shown in Figure 20-2) are common components of DSL's throughout much of the world ocean. The myctophids eat copepods (such as *Calanus finmarchicus*), and they in turn are eaten by squid, larger fish, and, in the polar regions, fur seals. This vertical food chain is illustrated in Figure 20-6.

DEFINITIONS

Adaptation. The process by which an organism becomes better suited to its environment.

Bioluminescence. Emission of light from living organisms.

Photophores. Specialized organs capable of bioluminescence.

Preadaptation. An adaptation for one environment or one aspect of that environment that is also suited for another environment or another aspect of that environment.

Vertical migration. An active (not passive) sustained movement up or down.

Exercise 20

Marine Adaptations and the Deep–Sea Environment

NAME _____

DATE _____

INSTRUCTOR _____

1. (a) Like other marine organisms, plankton display adaptations to their environment. Referring to Figure 20-1, list one morphologic adaptation for each of the following plankters that helps it to float or remain suspended.

Diatom _____

Flagellate _____

Copepod _____

(b) Why do the diatoms and flagellates need to stay at the surface, or at least in the shallow epipelagic zone? _____
_____
_____

(c) Why might a certain species of copepod need to stay at the surface or in the epipelagic zone? _____
_____
_____

2. For the following four species, list two adaptations for depth and give the function of each adaptation, specifying the condition of depth to which the adaptation evolved. Refer to Figure 20-1 and Table 20-1 for your answers.

	Adaptation for depth	Function
Lantern fish	_____	_____
	_____	_____
Anglerfish	_____	_____
	_____	_____
Swallowers	_____	_____
	_____	_____
Hatchetfish	_____	_____
	_____	_____

3. Brittle stars are a dominant organism in the abyssal benthic environments of the world ocean. As mentioned in the text, their main preadaptations are their long appendages and their ability to feed on detritus. Suggest why these two characteristics are especially good preadaptations for life in the deep-sea benthic environment. _____

_____

_____

_____

_____

4. After an examination of Table 20-1, where would you expect the greatest biomass of deep-sea benthos to be found?

    (a) Consider the environments below, and rank them by benthic biomass from greatest to least.
    Shallow benthic _____
    Deep benthic under upwelling _____
    Deep benthic not under upwelling _____
    (b) Explain your answer. _____

    _____

    _____

    _____

    _____

5. Answer the following questions on bioluminescence:

    (a) Some organisms squirt an ink that luminesces as it comes in contact with the dissolved oxygen in seawater. Suggest the purpose of this biochemical adaptation. _____

    _____

    (b) The photophores on a given fish species exhibit a species-specific pattern (that is, a pattern unique to that species). In fact, these patterns are so distinctive that fish taxonomists can use them to identify the various species of fish. Of what use might this specific photophore pattern be to the fish? _____

    _____

    (c) Not only does each species have a characteristic photophore pattern, but male and female members of the same species usually have slightly different patterns. Why? _____

    _____

    (d) On the lantern fish, photophores are found mainly in one region of the body (Figure 21-2). Suggest a reason for this particular concentration. (Note: The hatchetfish shown swimming under the lantern fish in Figure 20-1 are often found in this position, looking for the outlines of lantern fish so that they can attack them.) _____

    _____

    _____

    (e) There are probably many other reasons for photophores. Briefly answer the questions below.
        (i) Some anglerfish have a light organ (the only one they possess) at the end of their lure. Why? ____

        _____

        _____

        (ii) Some bottom-dwelling fish in the deep sea have a large aggregation of photophores around their heads and also very large eyes. Why? _____

        _____

        _____

6. Suggest uses for the following adaptions and the characteristic of the deep sea each is adapted for.

   (a) Lantern fish have scales, but many fish living in the deeper environments do not. Why? _____

_____

_____

   (b) Why do many deep-sea fish have poorly developed musculature? _____

_____

_____

   (c) Many deep-sea fish pass their juvenile stages in shallow water. What would be the advantages of this?

_____

_____

   (d) In many deep-sea species, the male becomes totally dependent on the female, and only its reproductive system continues to function fully. What is this an adaptation for, and why would you think it is a necessary adaptation for the deep-sea environment? _____

_____

_____

7. In phytoplankton vertical migration (Figure 20-3), the organisms cluster during the day in response to the light and disperse at night, in the absence of light.

   (a) What advantages do you think there might be in packing together during the day? _____

_____

_____

   (b) Why do the phytoplankton avoid the surface waters? _____

_____

_____

8. Recalling the trophic level of *Calanus finmarchicus*—its prey and predators—answer the following questions about its migratory behavior by referring to Figure 20-4:

   (a) Why do *Calanus finmarchicus* migrate to shallow water? _____

_____

_____

   (b) What is the most logical stimulus for this migration? In other words, how do these organisms know when to migrate upward? _____

_____

_____

   (c) Why are they in shallow water at night instead of in the daytime? _____

_____

_____

   (d) Why do they disperse at midnight, when all the food is in the upper 50 meters? _____

_____

_____

   (e) Why do the deeper forms not migrate up at dawn? _____

_____

(f) The migration pattern of *Calanus finmarchicus* is a behavioral adaptation that probably evolved over the course of a long geologic period, owing to natural selection. In a few well-reasoned sentences, state how this natural selection might have occurred such that a previous population of *Calanus finmarchicus* having equal numbers of members that (1) didn't migrate at all, _____

_____

_____

(2) migrated up in the daytime, and _____

_____

_____

(3) migrated up at night evolved into the type of population that we see today. _____

_____

_____

9. Answer the following questions about the deep scattering layers:
(a) In the recording of the DSL migration in Figure 20-5, three distinct layers can be seen (assume they are moving toward the surface as night approaches). There are probably many organisms in each layer, but name at least one organism that you think would be in the uppermost migrating layer.

_____

Why is this organism migrating to the surface? _____

_____

_____

Why do you suspect it is migrating at night instead of during the day? _____

_____

_____

What would be a logical reason for the organisms in the layers below to be migrating upward after the top layer? _____

_____

_____

Also, name at least one species that you would expect to find in the next layer. _____

_____

_____

(b) The migration pattern of most DSL's is diel (that is, they migrate daily), but that of the DSL's in the Arctic is not. What do you think the Arctic temporal migration pattern is, and why? _____

_____

_____

(c) DSL's are more highly developed (have a greater biomass and more layers) in some geographic areas than in others. Where would you expect to find the most highly developed DSL's, and why? Refer to the information in Table 20-1 if necessary, or to Exercise 15 on the distribution of life in the sea. _____

_____

_____

_____

_____

*E*stuaries

OBJECTIVES:

■ To understand the formation of various types of estuaries.

■ To understand the unique circulation and water-mixing patterns of estuaries.

■ To appreciate the biological and economic importance of estuaries.

An estuary is a semi-enclosed basin in which river water mixes with seawater. This meeting of water of different densities, combined with the effects of tides, gives rise to unique circulation patterns. Estuaries have a high nutrient content, and support rich biological communities. Many estuaries are heavily used by humans, and the management of conflicting uses is a growing concern.

The Formation of Estuaries

Estuaries are formed through a variety of geologic processes:

1. The drowning of river valleys when sea level rose at the end of the last glaciation of the Pleistocene Epoch. This type of estuary is common on the coastal plain of the United States.

2. The subsidence of fault blocks on tectonically active coastlines. South San Francisco Bay is an example.

3. The carving of fjords by glaciers. Naturally this type of estuary is found in high latitudes. The fjords of Norway, Greenland, and Alaska are some examples.

4. The development of sand spits and offshore bars along coastlines to form barrier beach-enclosed estuaries. These estuaries are common on the Atlantic and Gulf coasts of the United States. Pamlico Sound is an example of this type.

Estuarine Circulation

A distinguishing feature of estuaries is their unusual circulation, which is controlled by two major factors: water density and tides. First we will examine how density variations drive circulation within an estuary. Assuming no tidal effects, the continuous inflow of river water raises the local height of the water surface above that of the general ocean surface, creating a seaward slope as the river water flows out and over denser seawater. Consider the two "end-members" of the estuary (Figure 21-1a): a column of lower-density (i.e., lower-salinity) water at the river's mouth and a shorter column of higher-density seawater offshore. Because pressure at any depth is the product of the height of the overlying water column times the column's density, at some depth the pressure created by the freshwater column will equal that created by the shorter seawater column (i.e., the dashed line in Figure 21-1a). At any depth above this level, the pressure in the river-water column is greater than that in the seawater column, and water will therefore flow *seaward* along the horizontal pressure gradient. Conversely, at any depth below this level, pressure in the river-water column is less than that in the seawater column, and water will flow *landward*

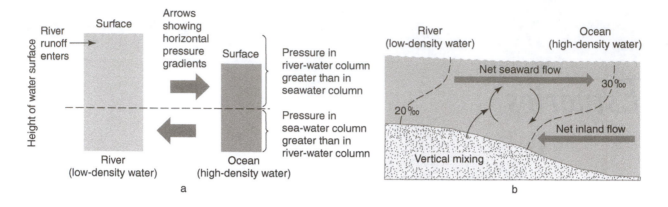

Figure 21-1 Density-driven circulation in estuaries. (a) Simple end-member model of river water and seawater columns showing change in direction of horizontal pressure gradients with depth. (b) Cross-section flow diagram, showing net-flow patterns, vertical mixing between fresh and salt water, and general salinity variations with 20‰ and 30‰ isohalines.

along the horizontal pressure gradient. Building from the principle illustrated by this two-column model, the more realistic and dynamic estuary cross section in Figure 21-1b illustrates the seaward flow of surface waters and inland flow of deeper waters. Vertical mixing between river water and seawater produces a net decrease in salinity landward, but note that isohalines (lines of equal salinity) bend sharply seaward toward the surface, reflecting the predominance of seawater at depth and fresh water near the surface.

Tides are the second major factor controlling estuarine circulation. Horizontal pressure gradients within the estuary shift in response to the periodic rise and fall of sea level driven by tidal forcing (Exercise 10). Thus, seawater will flow seaward during ebb tide and landward during flood tide. The seaward flow lasts longer and is of slightly greater velocity than the landward flow, resulting in a net seaward transport for a given tidal cycle. This back-and-forth "sloshing" of estuary waters in response to tides effectively mixes the water column vertically and also produces diffusion of seawater into the estuary.

The relative dominance of density versus tidal components determines an estuary's circulation type and salinity structure. Where tide-related mixing is weak, a sharp **halocline** (a zone in which salinity changes rapidly with depth) will separate the two well-mixed water layers. This circulation pattern is termed a *salt wedge* because of the

wedge-shaped intrusion of seawater into the estuary (Figure 21-2a). A classic example of a **salt-wedge estuary** is the lower Mississippi River, where tidal ranges are relatively low and river discharge volume is extremely large. At the other extreme of the estuary circulation spectrum is the **well-mixed estuary** (Figure 21-2b), in which tidal fluctuations effectively mix together river water and seawater, precluding large density differences within the estuary. Portions of Delaware Bay are good examples of the well-mixed estuary. Most estuaries, however, are intermediate between these two extremes and are classified as **partially mixed.** Whether density-driven or tidally driven circulation prevails depends upon the relative magnitudes of river discharge and tidal fluctuations. Since river discharge may vary considerably over the seasons, circulation characteristics may vary from season to season. In addition, circulation within relative large estuaries, such as Chesapeake Bay, may be influenced by the Coriolis effect (discussed in Exercise 9).

Biological and Economical Significance of Estuaries

Estuaries contain flourishing biological communities and are often referred to as the nurseries of the ocean. Their high nutrient content stimulates phytoplankton productivity (although light availability may limit this in many estuaries). These phyto-

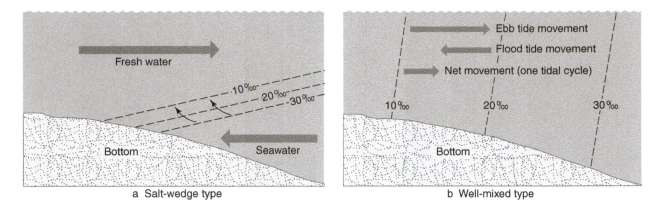

Figure 21-2 Comparison of interactions between river water and seawater, as reflected by salinity structure, in a salt-wedge estuary and a well-mixed estuary. (a) Within a salt-wedge estuary, the lower intrusion of seawater is reflected by the narrowly spaced, seaward-inclined isohalines. (b) Within a well-mixed estuary, mixing of fresh and salt water by tidal ebb and flood is reflected by the widely spaced, nearly vertical isohalines.

plankton in turn support a rich and diverse fauna—extensive shellfish beds may develop, many fish species use estuaries as a breeding ground, and tidal marshes along the estuarine shores support abundant waterfowl.

DEFINITIONS

Halocline. Zone in which salinity changes rapidly.

Partially mixed estuary. An estuary that shows a small to moderate salinity change with depth. Both density-driven and tidally driven circulation are important.

Salt-wedge estuary. An estuary in which density-driven circulation dominates and two well-mixed layers are separated by a sharp halocline. Seawater entering the estuary appears as a bottom tongue or wedge.

Well-mixed estuary. An estuary in which tidal fluctuations dominate, producing a well-mixed vertical water column.

Exercise *21*

Estuaries

NAME _____

DATE _____

INSTRUCTOR _____

Report

1. The surface salinity data in Table 21-1 were obtained from cruises up the Hudson River estuary during March and August of 1974. The bottom salinity during these months was approximately 1.2 times the surface salinity, indicating that the Hudson is a partially mixed estuary. In the table and map in Figure 21-3, "kp" is an abbreviation for kilometers north of the southern tip of Manhattan, and "kp−" means kilometers south of the same point. The isohaline (contour connecting points of equal salinity) is given for 1‰ only.

Using these data, plot the isohalines at 0.1‰, 10‰, 15‰, 20‰, and 25‰ for surface salinities during March and August. To do this, interpolate from the table to find the kp of the desired salinity. For example, to locate the 1‰ isohaline for the month of March—already plotted—you would interpolate between 0.5‰ (kp40) and 2‰ (kp23). Because 1‰ is roughly one-third the difference between 0.5 and 2‰, the desired kp is roughly one-third the distance (about 6 km) from kp 40 toward kp 23 (18 km). Thus, the 1‰ isohaline would be plotted at kp 34. Follow the same procedure for the month of August and you will find that the 1‰ isohaline should lie at about kp 76.

TABLE 21-1

1974 salinity data on the Hudson River estuary

	Surface salinity (‰)	
kp	**March**	**August**
108	< 0.1	< 0.1
85	< 0.1	0.5
66	< 0.1	1.5
55	< 0.1	4
40	0.5	7
23	2	12
2	9	19
−3	13	20
−13	16	23
−29	28	30

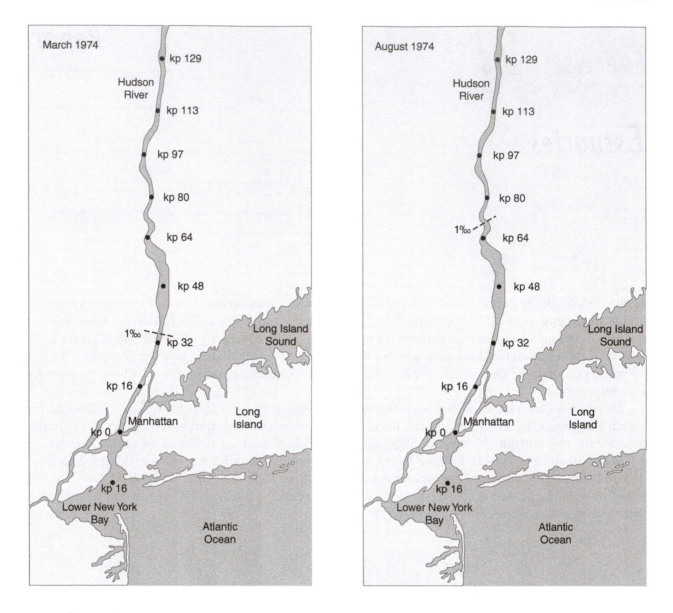

Figure 21-3 Surface salinity map of the Hudson River and its estuary, with the contour given for 1‰.

2. Appreciate that the data in Table 21-1 are time-averaged salinities. The salinity at kp 32, for example, will change with the phase of the tide, increasing as the tide floods and decreasing as the tide ebbs. Answer the following questions about the isohaline contours you plotted in Question 1.

(a) During what month does salt water extend the farthest up the estuary? _____

(b) From the principles discussed in the text, hypothesize what might cause this variation. _____
_____
_____
_____

(c) The average movement of water in a tidal cycle is 13 km. If the time-averaged surface salinity at kp 16 is 9‰ in August, calculate the maximum and minimum salinities over a tidal cycle at this point. (Hint: Between mid-tide and high tide, water will move 6 km upstream.)

Maximum kp 32 salinity _____‰   Minimum kp 32 salinity _____‰

| TABLE 21-2 | | |

Freshwater flow at Green Island Dam, kp 248

| | Flow | |
| | (cubic feet | (cubic meters |
Water year 1974	per second)	per second)
October 1973	4000	145
November	9000	330
December	24000	870
January 1974	17000	620
February	17000	620
March	20000	730
April	30000	1100
May	24000	870
June	10000	365
July	11000	400
August	7000	255
September	8000	290

3. The seasonal precipitation pattern in the Hudson drainage basin is quite consistent from year to year. The last point on the Hudson at which river flow is monitored is the Green Island Dam at kp 248. The freshwater flow at this dam, averaged over one-month intervals, is shown in Table 21-2.

   (a) On the graph below, plot the monthly average flow as a bar graph.

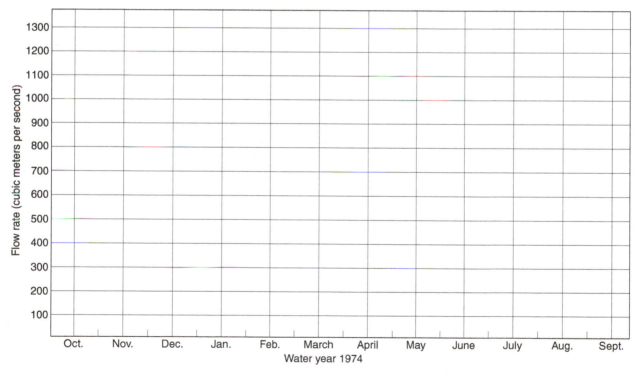

   (b) Why is there maximum flow during March through May? _____

_____

_____

_____

(c) Why is there minimum flow during August through October? _____

_____

_____

_____

4. On average, each person in the New York metropolitan region uses more than 630 liters of water per day. The total demand is nearly 9×10^9 liters per day, much of which is met by tapping the waters in the Catskill Mountains west of the Hudson. In dry years, precipitation within the Catskills is not sufficient to meet this demand, and the Hudson is the only alternative water source. To maintain water quality, it is desirable to draw Hudson water with a salinity of less than 0.1‰. Note that the position of the 0.1‰ isohaline is linearly related to the freshwater flow rate over Green Island Dam. Thus, it would be important

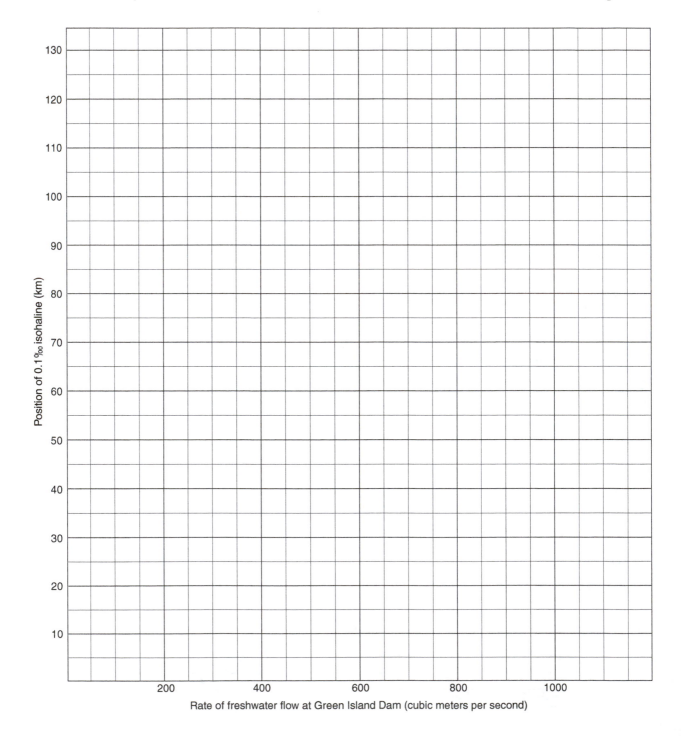

Rate of freshwater flow at Green Island Dam (cubic meters per second)

to determine the optimal kp at which to place a water intake pipe such that < 0.1‰ waters could be drawn in a drought that reduced Green Island Dam flow to 100 cubic meters per second.

(a) To determine the ideal kp location for a water intake pipe for both March and August, first plot the monthly flow rate over Green Island Dam versus the kp of the 0.1‰ isohaline for the same month.

(b) From your plotted data, decide where you would install the water intake pipe. Appreciate that the farther north you install the intake, the more expensive it will be to pipe the water to New York City. Note that the present water intake is at kp 108. _____

_____

5. Most of the sewage generated by New York City and northern New Jersey is discharged into the lower Hudson after treatment. Is it possible that these discharges may affect water quality in upstream regions to the north of New York City? Explain why or why not. _____

_____

_____

_____

_____

_____

6. A critical aspect of estuary pollution management is understanding how quickly the system completely "flushes," or replaces, the water contained within it. However, precise and accurate calculation of an estuary's flushing time is difficult; variables that must be considered include the basin volume, river input, tidal input and periodicity, physical and chemical differences between the ocean and river water, and the nature of the pollutant. Outlined below is a numerical method to determine how pollutant concentration will decrease as tidal cycles flush the estuary. For this calculation, an estuary is divided into two volumes: the upper A volume (mixed river and seawater) and the lower B volume (seawater). Inputs into the estuary are the net river water input (Q_R) and the ocean water input (Q_B). Output from the estuary is in the form of surface flow to the ocean (Q_A). The fundamental assumption is that any pollutant mixes rapidly within the lower B volume, mixes upward gradually, and exits with the surface waters. The numerical relationships are as follows:

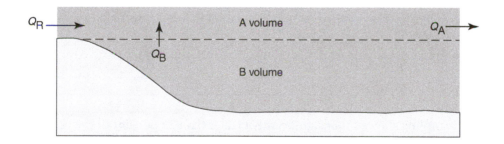

$Q_A = Q_B + Q_R$ [average outflow is equal to average inflow]

$f_A = Q_A/A$ [fraction of A lost to ocean during tidal cycle]

$f_B = Q_B/B$ [fraction of B passing into A during tidal cycle]

$f_X = (f_A + f_B)/2$ [mean exchange ratio for the estuary]

$C_t = A(1 - f_A)^t + (t f_X/(1 - f_X))(B(1 - f_X)^t)$ [volume of contaminant C left after t tidal cycles]

$\%C_t = (C_t/C_o)100$ [% contaminant remaining after t tidal cycles]

Using the above relationships and the values below for a given estuary, calculate the percentage of a contaminant remaining after 25, 50, 75, 100, 125, and 150 tidal cycles. Plot these data points and connect them with a line on the graph below. Use the values below to solve the first four equations, then use these calculated values in the last two equations to determine the percentage of the contaminant remaining after each tidal cycle.

$$A = 10 \text{ units} \qquad B = 15 \text{ units} \qquad Q_B = 0.4 \qquad Q_R = 0.1$$

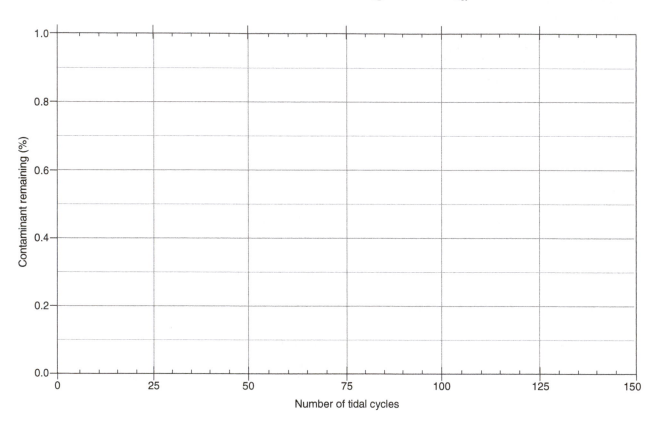

(a) From this model, how many tidal cycles are required to reduce the original pollutant concentration by 80 percent? _____

(b) Given that there are roughly 60 tidal cycles per month, about how many days does this represent? _____
_____

(c) If estuary sampling showed that significantly greater than 20 percent of the pollutant still remained after your estimated flushing time above, discuss two possible explanations for this discrepancy.
_____
_____
_____
_____

(d) If estuary sampling showed that significantly less than 20 percent of the pollutant remained after your estimated flushing time above, discuss two possible explanations for this discrepancy. _____
_____
_____
_____
_____

Littoral Ecosystems and Environmental Impacts

OBJECTIVES:

■ To examine the accessible shore environment, its zonation, and the invertebrates that characterize each zone.

■ To introduce coral reef environments and illustrate their fragility.

■ To study the impact on marine life of actual oil spills off the Texas coast.

*L*ittoral environments are those found from the shoreline to a depth where fixed benthic plants can no longer exist. They are divided into the *seawater spray zone* above the highest tide level, the *supralittoral zone,* the *littoral* (intertidal) *zone* between low and high tide extremes, and the *sublittoral zone,* which extends seaward from the lowest low tide level to a water depth that can no longer support fixed plants (see Exercise 15). These environments are easily accessible to humans, and the data for this exercise were collected by students like yourselves. They are also especially vulnerable to both natural and human disturbances, and we will look at the effects of such disturbances on these environments. We will begin in temperate climates with rocky environments, sand beaches, and estuarine environments, and then travel to the tropics to study South Pacific coral reef environments.

Temperate Rocky Environments

Rocky environments are common in the New England states and along the west coast of the United States. However, any hard surface can constitute this type of environment, and some of the examples used here are actually from pier pilings at Pacific Beach in San Diego, California.

In the seawater spray zone there is found — but often overlooked — a band of black crust looking as though it were painted on the rocks. This band marks the region between land and sea environments, and here we find blue-green (cyanobacteria) and green algae, bacteria, and small lichens. Below this zone, the top of the littoral zone is marked by the first occurrence of barnacles, usually small species called buckshot barnacles. Barnacles are crustaceans related to crabs and lobsters, and in general their size increases with depth. Most barnacles are hermaphroditic (both sexes), and they fertilize adjacent barnacles. The eggs develop inside the barnacle until the planktonic larvae, known as *nauplii* (see Exercise 19), are discharged into the seawater, where they may spend hours searching for the right rock surface to settle upon. The nauplius settles on its head, where an organ secretes a glue to attach it to the rock. The barnacle exists in this upside-down position for the rest of its life, straining seawater through its modified legs to catch plankton to eat.

The greatest density of organisms in the littoral zone is found at mid-tide level. In California, this area is occupied by the "big three" — the gooseneck barnacle, the black mussel, and the starfish, *Pisaster.* The barnacles are stuck forever in one spot, whereas the small black mussels can move about slowly within the zone by detaching and reattaching their mooring lines, called byssal threads. The starfish is very mobile and searches for mussels, which it pulls open using suckers on its tube feet, everting its stomach into the mussel to digest its soft parts.

The deepest part of the littoral zone is the algal-mat zone, which is almost continuously covered by water, allowing thick layers of algae to form. On the east coast of the United States, some rocky intertidal zones exhibit algal mats starting with brown algae at the top and grading downward to red algae.

Temperate Sublittoral Environments

In the sublittoral zone of southern California, large "forests" of the brown alga *Macrocystis* are found. Commonly known as kelp, it is the largest plant in the world, and fastens itself to rocks with a root-like growth called a holdfast. The plant may grow up to 2 feet per day, making kelp beds one of the highest productivity regions in the world. Kelp is harvested by large ships and processed on land into commodities useful in the chemical and food industries. Inexpensive ice cream, for example, is thickened with a kelp by-product.

In early 1992, a sewage pipe carrying 180 million gallons per day of advanced primary-treated sewage ruptured at a depth of about 10 meters in the middle of the Point Loma kelp beds. For about two months a sewage boil about the size of a football field existed at the surface within the kelp bed. Sampling and observations during the spill indicated that the surface waters at the spill site were discolored and bad-smelling. The mixing of sewage effluent and seawater reduced salinity around the rupture to about one-half the normal level (about 15‰). The sediment load of the sewage coated the blades of the giant kelp, causing it to become limp and withdrawn from the surface. Kelp at the edges of the spill were healthy and dense, more so than normal.

Temperate Sand Beach Environments

On November 1, 1979, the oil tanker *Burmah Agate* was involved in a collision with another ship about 7 kilometers outside the Galveston, Texas, break-water, and sank in water 13 meters deep (Figure 22-1). The tanker spilled crude oil that burned for

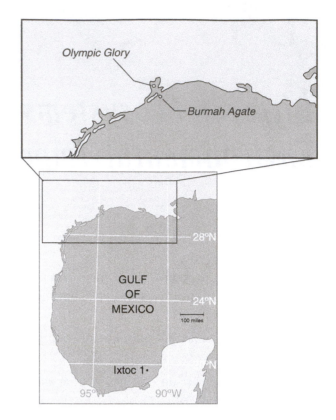

***Figure* 22-1** Map showing the locations of the oil spills discussed in this exercise. The *Burmah Agate* and *Olympic Glory* were both tankers that spilled oil off the coast of Galveston, Texas, and in the upper reaches of Galveston Bay, respectively. [From R. Casey et al., *Proceedings of the Offshore Technology Conference,* 1982, pp. 449–459.]

69 days. The day after the spill started, planktic and benthic surveys were made near the burning vessel, on one transect south of the ship and one to within 200 meters of the beach at Galveston along the predicted oil-spill track. The oil did in fact take the predicted track to Galveston Beach, and sampling was performed at 1 week, 1 month, 6 months, 1 year, 1.5 years, and 2 years after the spill (Figure 22-2). The impact and recovery data from the spill are presented in Table 22-1.

Temperate Estuarine Environments

In 1981, the tanker *Olympic Glory* was involved in an accident and spilled oil in upper Galveston Bay, Texas (location 1 in Figure 22-3) The ship was maneuvered to Barbour's Cut (location 2), where most of the released oil was contained. Samples of

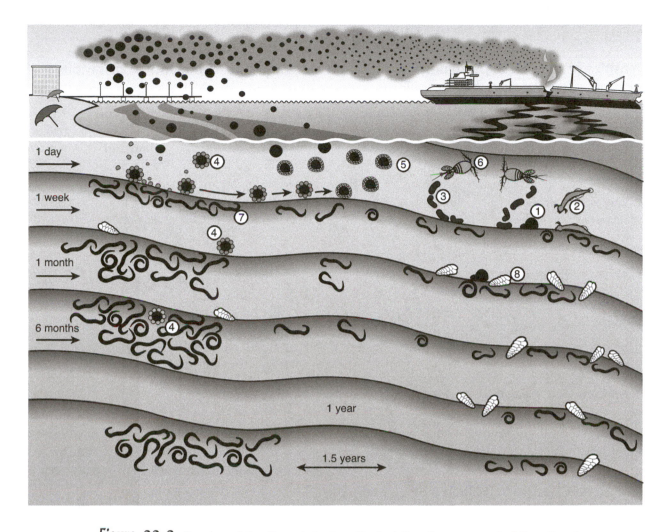

Figure 22-2 Drawing of the *Burmah Agate* spill, depicting the movement of the oil and suspended sediments and their effects on organisms. The *Burmah Agate* is shown at the upper right, spilling oil that moves on the ocean surface toward the Galveston beach. Oil drops in the smoke from the burning oil are also carried toward the beach. From the Galveston breakwater at the upper left, a plume of sediment- (clay-) laden water moves parallel to the Galveston beach. Nearshore (off the beach) waves stir up silt and suspend it throughout the nearshore waters. Findings taken 1 day, 1 week, 1 month, 6 months, 1 year, and 1.5 years after the spill are shown. 1 denotes oil balls; 2, recently dead chaetognaths; 3, copepod fecal pellets; 4, oil drops coated with sand; 5, oil and water spheres coated with silt; 6, copepods feeding on oil; 7, nematodes; and 8, the benthic forminiferan *Brizalina lowmani.*

plankton and benthon were taken for a year from a fishing pier at Sylvan Beach, which is near the original accident site and a few miles south of Barbour's Cut. The results are presented in Table 22-2. The most important biological changes off Sylvan Beach are shown in Figure 22-3.

Oil spills have a negative impact on sea life, and every effort should be made to avoid them. In the two examples presented here, considerable effort was made to minimize the effects of the oil on marine life. The *Olympic Glory* was taken into Barbour's Cut and oil curtains were deployed, forming a barrier to floating oil. Oil curtains were also deployed for the *Burmah Agate,* and oil skimmers were used to remove oil from the sea surface. Commonly some absorbant, such as hay, is spread on beaches that are in

TABLE 22-1

Densities of organisms (in number per 10 cubic centimeters) in sediments from various locations after the *Burmah Agate* spill*

	1 day (early winter)	1 month (winter)	6 months (spring)	1 year (early winter)	1.5 years (spring)	2 years (early winter)
Nearshore						
Benthic foraminiferans	40	300	60	20	0	20
Nematodes	1300	200	320	180	480	340
Near ship						
Benthic foraminiferans	8	24	30	22	30	30
Nematodes	22	56	120	48	260	90
Offshore						
Benthic foraminiferans	20	30	0	30	60	0
Nematodes	40	0	100	120	310	220

* *Source:* R. Casey et al., Proceedings of the Offshore Technology Conference, 1982.

the path of an oil spill. Should the oil be deposited on the beach, the oil-soaked hay can be removed.

Coral Reef Environments

On January 3, 1993, Cyclone Kina hit the Fiji Islands and caused heavy damage. As part of a long-term global change survey, the Fiji reef system was studied by biologists first in 1992, and again in the year following the storm. Two reef areas were selected for detailed study; the first was off the Savusavu Airport on Vanua Levu Island, Fiji, and the second was about 15 kilometers to the east of the airport at the Mu Mu Resort (Figure 22-4). The

TABLE 22-2

Standing crops of microbenthon and microplankton from sediment and water samples, respectively, at Sylvan Beach after *Olympic Glory* spill*

	2/26/81	3/5/81	3/21/81	3/27/81	3/31/81	5/23/81	1/11/82	1/24/82	1/26/82†	1/26/82†	2/1/82	2/5/82
Sediment samples												
Benthic foraminiferans	290	410	2540	1100	nc	1800	200	nc	nc	nc	nc	nc
Nematodes	300	1000	600	980	nc	250	290	nc	nc	nc	nc	nc
Water samples												
Diroflagellates	16	23	48	6	0.1	0.3	0.07	11.5	4.2	2.3	1.4	0.3
Diatoms	0.1	17	0.05	0.02	0.4	0.06	0.07	1.3	1.9	1.0	0.3	0.5

Source: R. Casey et al., *Proceedings of the Offshore Technology Conference,* 1982.

* Number of benthic foraminiferans and nematodes per 10 cubic centimeters from sediment samples; millions of dinoflagellates and diatoms per liter from water samples. Sampling period: 1 year; nc means not counted.

† Sampling was done at 0900 and at 1430 hours on this date.

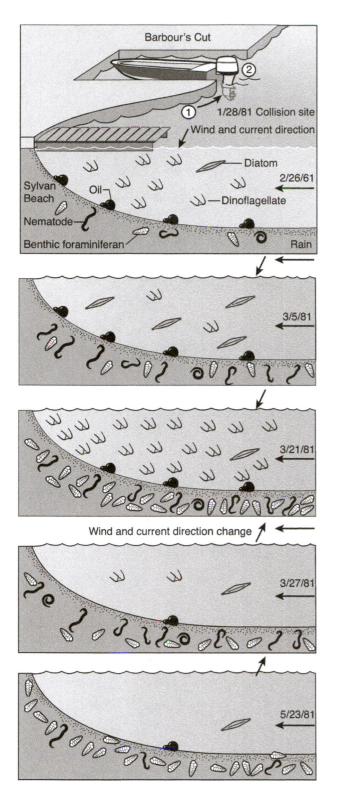

Figure **22-3** The *Olympic Glory* oil spill and its observed effects on microplankton and microbenthon. Shown are the collision site (1) and the location of the leaking tanker after the spill (2). Below the sketch of the fishing pier from which the samples were taken are drawings depicting the oil, microplankton, and microbenthon in a time series from 1 month until about 4 months after the spill. Important meterorological and oceanographic events—rain, wind, and current direction change—are also shown. Diatoms and dinoflagellates in the water column and nematodes, benthic foraminiferans, and oil on the bottom are shown in their relative abundances at intervals after the spill. That is, four diatoms drawn means that twice as many diatoms were collected relative to a collection time for which only two diatoms are drawn. The figure does not imply the actual density of the organism, nor the relative densities of different groups in the same diagram.

then observed at greater intervals out to the reef crest several hundred meters offshore.

The main changes on the Airport transect between 1992 and 1994 are shown in Figure 22-5. The inner reef flat corresponds to that portion of the reef that was documented at 20-centimeter

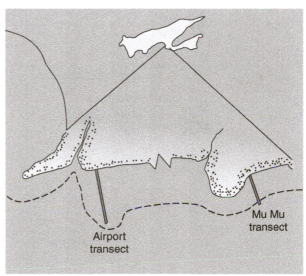

reef organisms were observed using a box with a glass bottom (a view box) along each transect at 20-centimeter intervals out to a distance of 45 meters at the Airport transect, and out to 25 meters offshore at the Mu Mu transect. The reefs were

Figure **22-4** Sketch map of coral transect locations off Vanua Levu in the Fiji Islands, indicating the Airport transect (left) and the Mu Mu transect (right). Dashed line indicates seaward end of the reef crest. Note the small submarine canyon just west of the Airport transect.

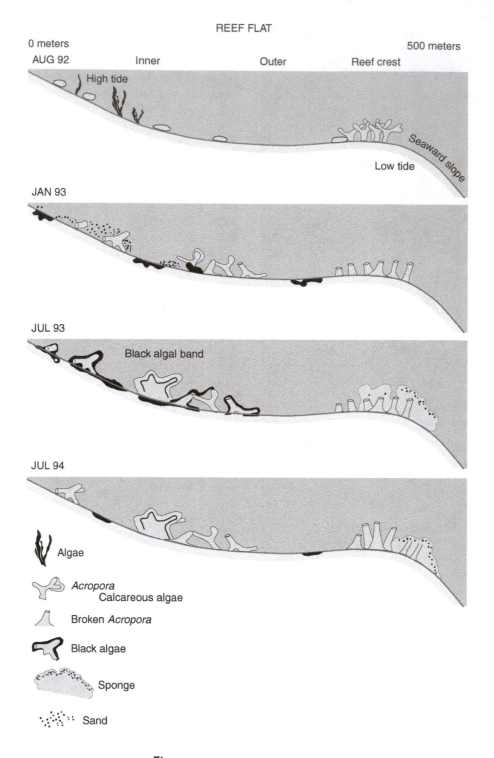

REEF FLAT

0 meters   500 meters
AUG 92   Inner   Outer   Reef crest

High tide

Low tide
Seaward slope

JAN 93

JUL 93

Black algal band

JUL 94

ΨΨ Algae

Acropora
Calcareous algae

Broken Acropora

Black algae

Sponge

Sand

Figure 22-5 The Airport transect.

intervals. The outer reef flat extends out 500 meters to the reef crest, which is slightly elevated above the flat and is dominated by the staghorn coral *Acropora*. One can see that Cyclone Kina destroyed about 80 percent of the *Acropora* at the Airport transect, and that the coral debris was transported shoreward, filling depressions on the flat. This coral was not bleached like old reef debris, indicating its recent origin, and in some cases this coral "hash" may have been transported 400 meters over the reef flat by wave action. Large numbers of dead sea urchins were found with broken and missing spines. Starfish seemed unaffected, probably owing to their ability to hide in crevices in the reef flat. By July 1993, the outer portion of the inner reef flat was covered by black encrusting algae forming a black band. This algal band appears to have covered a 10-meter-wide patch about 80 meters from shore. The band appears to have covered some of the broken staghorn coral. In the mid-reef area the reef rubble, including coral hash, was mainly encrusted with pink calcareous algae. A layer of brown sponge covered the cyclone-cropped *Acropora* stands on the outer reef area.

The reef along the Mu Mu Resort transect appeared less damaged than the Airport reef (Figure 22-6). There was significantly less coral hash on the reef flat, and the stands of living *Acropora* were not planed off, as were those along the Airport transect. There was no significant change between the transect in July and the one just after the storm in January 1993. The only similarity to the Airport transect was a band of black encrusting algae that covered a 10-meter wide patch of coral debris about 70 meters from shore.

Coral Reefs, El Niño, and Cyclones

Very warm water has an adverse effect on coral. Warm-water episodes typify El Niño conditions in the western Pacific, which lead to coral damage and subsequent bleaching of the dead coral. The reef damage at Fiji was a one-two punch—first the weakening of the coral by 3 years of warm El Niño waters; then the knock-out punch of Cyclone Kina.

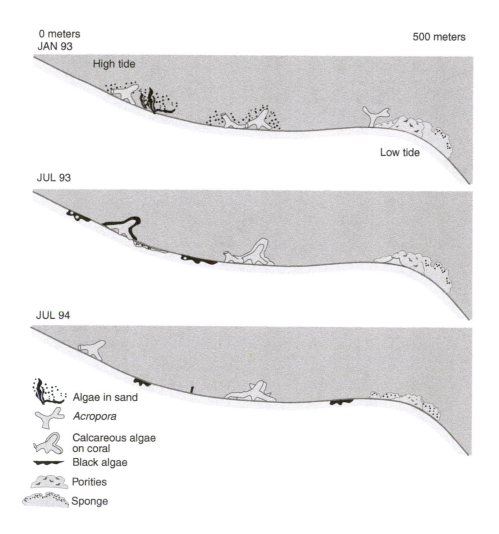

Figure 22-6 The Mu Mu Resort transect.

Black band disease is one of many diseases that attack damaged or weakened coral, and it was very obvious at both sites. The study suggests that the transects could have been located anywhere around the island and the results would have been the same: damage to branching *Acropora* coral and transportation of the coral debris onto the reef flat, with black algae covering the dead coral in linear bands. It is anticipated that the reef will return to its precyclone condition within the next 30 years. However, the impact of the great El Niño of 1998 has yet to be evaluated.

Exercise 22

Littoral Ecosystems and Environmental Impacts

NAME _____

DATE _____

INSTRUCTOR _____

1. Using the illustrations provided below, separate the upper intertidal from the mid-tidal environment with a line on the picture, then label separate pictures of each (inter & mid-tidal).

2. A cluster of mid-tidal organisms (1 middle picture) occurs in the supralittoral zone. How might this happen?

_____

_____

_____

_____

_____

3. Why do you think barnacles deeper in the littoral zone are larger than those found at the top of that zone?

_____

_____

_____

_____

_____

_____

4. The Airport coral reef in Fiji was damaged more by Cyclone Kina than was the Mu Mu reef. Both areas were equally exposed to the ravages of the cyclone. What oceanographic condition at the Airport reef might have caused greater damage there? _____

_____

_____

_____

_____

_____

5. At the *Burmah Agate* spill site on the day after the collision ("1 Day" in Figure 22-2), tar balls were found on the bottom, as were soft-bodied plankton such as chaetognaths and abundant copepod fecal pellets (see Figure 22-2 and Table 22-1). However, in the water column above, copepods were the only abundant plankton. Apparently some plankton in the spill area were adversely affected and some were not. Study the figure and cite two pieces of evidence that suggest that at least one plankter was not immediately adversely affected by the spill. _____

_____

_____

_____

_____

_____

6. On the sand bottom near shore, small oil drops (20 micrometers) were found coated with silt-sized sediment (4 in Figure 22-2). On a mud bottom at the middle of the Galveston transect, there were found larger spheres (80 micrometers) of oil plus water covered with clay-sized particles (5 in Figure 22-2). Describe the most likely routes of the oil drops and spheres to the bottom. That is, what different sequences of events might you think of that formed small oil drops or spheres that later sank to the bottom? (Keep in mind the following: oil floats on water; a plume of clay-laden water exited the Galveston breakwater; wave action at the shoreline suspended silt-sized particles throughout the water column; and how the oil got to this region.) _____

_____

_____

_____

_____

_____

_____

7. *Olympic Glory* spill data for the Sylvan Beach area are given in Table 22-2. Samples were taken at intervals beginning a month after the spill and ending about a year after it. Plot the densities specified in parts (a) and (b) on the accompanying graphs.

   (a) Densities of benthic foraminiferans and nematodes from the sediment. Use solid lines to connect the data points for foraminiferans and dashed lines for nematodes.

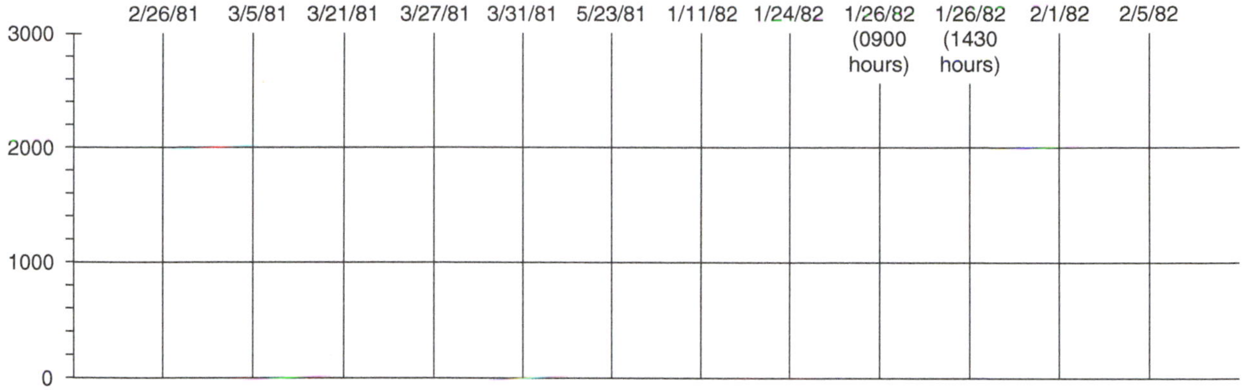

   (b) Densities of dinoflagellates and diatoms from the water column. Use solid lines to connect the data points for dinoflagellates and dashed lines for diatoms.

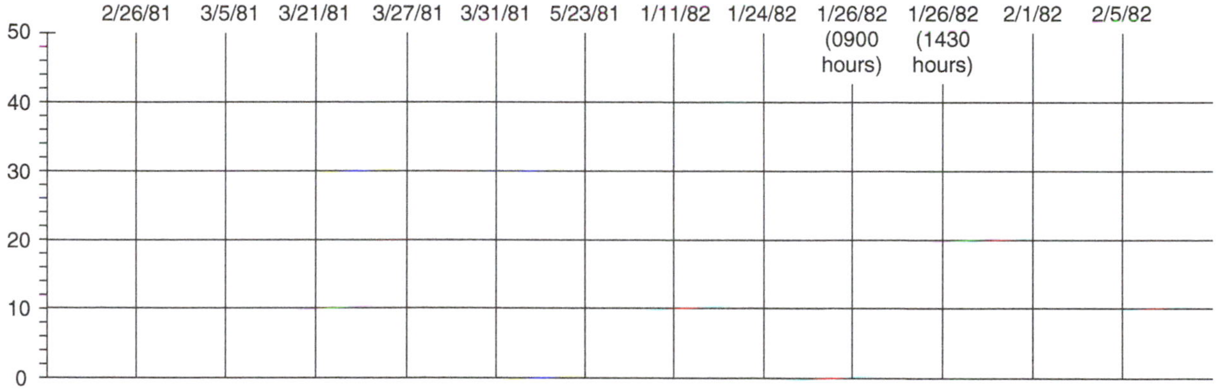

8. Red tide conditions (dinoflagellate densities greater than 1×10^6/liter of seawater) existed off Sylvan Beach for about a month after the *Olympic Glory* oil spill, until March 27, 1981. Can you think of any way that the oil spill might have contributed to these prolonged dinoflagellate blooms, which are not natural here at this time of the year? (Red tides normally occur here around the end of January and last for about a week.) _____

_____

_____

_____

_____

_____

_____

_____

_____

_____

_____

9. The red tide sampled at Sylvan Beach after the oil spill on March 5 was a diatom-dinoflagellate bloom rather than dinoflagellates only. By March 21 it had reverted to a normal dinoflagellate red tide (Table 22-2).

(a) How might these observations be accounted for? (Hints: What are diatom skeletons composed of, and what happened between February 26 and March 5, 1981?) _____

_____

_____

_____

_____

_____

_____

_____

_____

_____

_____

(b) From the information you have been asked to evaluate here and from what you have learned in lectures and the laboratory, what are the general oceanographic conditions that tend to cause natural red tides at Sylvan Beach, and why? _____

_____

_____

_____

_____

_____

_____

_____

_____

_____

_____

_____

*M*icropaleontology and Paleoceanography

OBJECTIVES:

■ To understand how microfossils are used to construct relative time scales.

■ To reconstruct past environmental conditions and changes using microfossils.

Paleoceanography is the reconstruction of the ocean's history and its past interactions with the atmosphere, biosphere, and lithosphere. The major source of paleoceanographic data comes from the coring of ancient sediments that have accumulated in ocean basins. Since the 1960s, an international consortium of scientists have worked together to collect such cores from the deep sea. This project was first known as the Deep Sea Drilling Project, and its specialized drill ship was the *Glomar Challenger*. The project was later renamed the Ocean Drilling Program and currently collects paleoceanographic data using the drill ship *Joides Resolution*. On board such drill ships, micropaleontologists, sedimentologists, petrologists, and other scientists perform preliminary studies as cores are collected to guide the ongoing drilling operations. After the cores are returned to land, subsamples are made available to scientists all over the world, who study the cores' historical record of events. Microfossils, or the skeletal remains of tiny organisms such as foraminiferans, diatoms, radiolarians, and coccolithophores (see Exercise 5), are extremely important in paleoceanographic research for two reasons: (1) their utility in dating and correlating core sediments from different areas, and (2) their recording of ancient oceanographic conditions

that prevailed in a given region during their lifetime. In this exercise, microfossil data will be used to explore both of these topics.

Using Microfossils to Determine Relative Ages of Sediments

Since the origin of life, innumerable species have evolved and become extinct on planet earth. Upon their deaths, the skeletons of individual members of a species may be incorporated within accumulating sediments and preserved as **fossils.** With respect to a given fossil species, all time can be divided into three parts: (1) the time interval *prior* to its evolution, (2) the finite time interval *during* its existence, and (3) the time interval *after* its extinction. The interval of time when a species lived on the planet is defined as its **range.** One can use the appearances and disappearances of different fossil species as one moves up the rock record through sequentially deposited (i.e., younger) layers to construct a **relative time scale;** this technique is known as **biostratigraphy.** Note that this approach provides a *relative* age, which differs from absolute age in years as determined through radioisotopes and their exponential decay.

Biostratigraphy is not an exercise in circular reasoning. It operates on two logical principles: (1) younger sediments are deposited on top of older sediments (i.e., you can't easily wedge fresh sediments between existing sedimentary layers), and (2) while particular species were living on the earth, some proportion of their skeletons were preserved in the sediment accumulating at the time. The first principle is analogous to putting your newspaper into a recycling bin at the end of each day: the oldest newspaper is on the bottom, and progressively younger newpapers accumulate on top. Now imagine that for some reason, no specific dates were ever mentioned in newpapers, but they were always piled sequentially in their bin. Although there are no actual dates, a series of overlapping unique events is recorded vertically in the bin in the form of "news." For example, the *Apollo* moon landing and the *Challenger* shuttle explosion were each unique historical events and were extensively recorded in all newspapers of their time. Thus, wherever bins of old, undated newspapers spanning both events are preserved, the moon landing would always be found lower in the bin than the *Challenger* disaster. If you were to examine a small bin in which the *Apollo* landing dominated the news, you would infer that the newspapers in that bin were relatively older than those in a small bin in which the *Challenger* explosion dominated the news. In this analogy, the bins represent various outcrops and cores of the earth's sedimentary layers, and the unique news events represent the second biostratigraphic principle: the recording of the relative occurrence of individual species on the planet. Thus,

using the vertical distributions of news events (i.e., species) from various bins (i.e., sedimentary outcrops and cores), one can reconstruct a relative time scale of historical events since the time that the newspapers (i.e., sediments) started accumulating.

Appreciate the importance of being able to order and correlate events recorded within different cores. If this were not possible, then we could not reconstruct the timing of events in the geologic past. Hence, any attempt to locate natural resources or reconstruct ancient history would be equivalent to poking around in the dark with a stick: anything you might discover would be an isolated, random find with no larger framework. Thus, the goal of biostratigraphy is to date rock layers as finely as possible and correlate them over as wide a geographic area as possible. For example, if you were drilling for oil and discovered that regional rocks of Late Miocene age are particularly rich in petroleum, then it would be extremely useful to know whether a given core was "too shallow in depth/too late in time" or "too deep in depth/too early in time."

Not all fossil species are good for chronologically ordering rocks and sediments. For example,

TABLE 23-1

Hypothetical composite range chart for five radiolarian species.

Epoch (time)		Approximate age in millions of years	Hypothetical radiolarian species				
			A	B	C	D	E
Pleistocene 1.6		1					
Pliocene 3.4	Late	2					
		3					
	Early	4					
5.3		5					
		6					
		7					
	Late	8		B			
		9					
		10					
Miocene 11.2		11				D	
		12					
		13					
	Middle	14					
		15					
		16					
16.6	Early	17	A				
					C		E

one that ranged from the Cambrian Period to today and lived only in a very restricted environment—say, the tropical intertidal zone—would be of little use for relative dating and correlation between different cores and rock types. The best fossil species for biostratigraphy are called **index fossils** and exhibit the following characteristics: (1) cosmopolitan distribution, (2) broad environmental tolerance, (3) high abundance, (4) short geologic range, and (5) distinct morphology. As you can imagine, a good index fossil would be one that lived over the entire surface of an ancient ocean for a short interval of geologic time. As individuals died and sank, their skeletons would be incorporated into the bottom sediments, which might vary from nearshore sand to deep-sea ooze. Finding individuals of an index fossil preserved in rocks from different ancient environments would allow you to correlate them as being deposited during the same time period. Revisiting our newspaper analogy, index fossils would be the equivalent of major news stories recorded by all countries (e.g., the fall of the former Soviet Union), while poor index fossils would be analogous to events restricted to local newpapers (e.g., "Freddy Jones wins the Pleasantville spelling bee") or events covered in newspapers with a very low, but perhaps widespread, abundance (e.g., "Bee Keeper's Daily").

In the following exercises you will "do" biostratigraphy by compiling the relative ranges of different fossil species from different cores into a composite range chart. This composite range chart can then be used to determine the relative ages of samples from new cores. An example of a composite range chart for radiolarians is given in Table 23-1. Note that this process is exactly how the modern geologic time scale (Appendix B) was constructed, albeit on a much larger scale and over a longer period of time. Also, appreciate that composite range charts represent a hypothesis about the relative age relationships of different fossil species; this hypothesis is tested each and every time the relative occurrences of fossil species in a new core are compared to, and incorporated within the composite range chart.

Reconstruction of Paleoceanographic Environments

To reconstruct paleoceanographic conditions using microfossil data from cores, we must correlate par-

ticular microfossils with the environmental conditions in which they lived. If living descendants of these microfossils can be correlated with specific environmental parameters (such as temperature, nutrient requirements, etc.), it can be assumed with some confidence that the presence of fossil relatives in a core interval indicates that similar conditions existed during the deposition of that core interval on the ancient sea floor. This use of the "present as the key to the past" is known as *uniformitarianism.* To demonstrate the uniformitarian approach, modern radiolarians are discussed and will serve as a guide for interpreting ancient samples in the exercises.

Radiolarians are planktonic protozoans with siliceous skeletons about 100–200 micrometers in diameter. Of the 500 or so living radiolarian species in the ocean today, about 250 species live at epipelagic depths in tropical to temperature latitudes, about 50 species live at epipelagic depths in polar latitudes, and about 200 live at mesopelagic or greater depths. A summary of some common radiolarian species, their geologic ranges, and their environments is shown in Figure 23-1. The photomicrographs of radiolarian skeletons in Figures 23-2–23-4 were collected and concentrated from the sea floor at various locations. Figure 23-2 is from the tropical Pacific (close to area C in Color Plate 8) and shows a great diversity of forms, including many restricted to warm waters. Figure 23-3 is from the polar Pacific (close to area D in Color Plate 8) and shows less diversity than Figure 23-2; some of the radiolarian species present are restricted to cold polar waters, and many diatoms are present.

The three photomicrographs in Figure 23-4 represent samples collected from temperate waters of the North Pacific, but from different productivity regimes. Figure 23-4a is from an open-ocean, low-productivity (oligotrophic) region in the middle of the North Pacific gyre (close to area A in Color Plate 8), whereas Figures 23-4b and 23-4c were collected from the nearshore, high-productivity (eutrophic) California Current region (close to area B in Color Plate 8). As shown by the latter two samples, the polar-derived California Current contains relatively more cold-water radiolarians, but is less diverse than the oligotrophic sample from the North Pacific gyre. Present in all the samples are plectopyramid radiolarians that dwell at mesopelagic and greater depths where physical and chemical conditions are very uniform, especially in the bathypelagic. The uniform, widespread

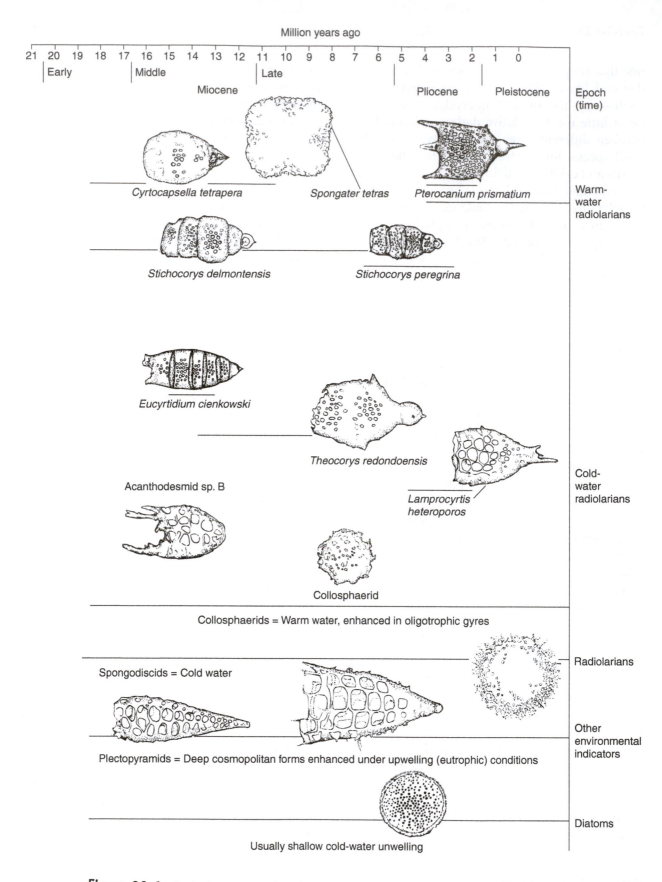

Figure 23-1 Geologic ranges and environmental distributions of some common radiolarians.

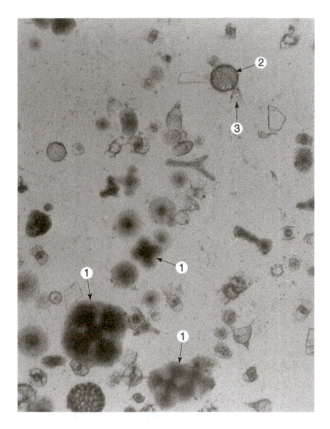

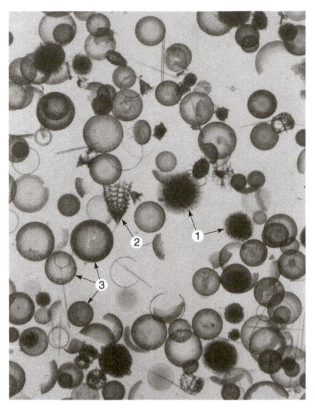

Figure 23-2 A high-diversity radiolarian assemblage from the warm tropical Pacific. The shallow-dwelling *Spongater tetras* (1) and collosphaerids (2) are common, and a deep-dwelling plectopyramid (3) is also present.

Figure 23-3 A low-diversity radiolarian assemblage from the cold polar Pacific. The assemblage is dominated by cold-water spongodiscids (1) and deep-water plectopyramids (2). Notice that diatoms (3), another microorganism predominant in cold surface waters, constitute a major portion of the sample.

environment at greater depth allows radiolarians and other plankton adapted to such conditions to be widely distributed from polar to tropical latitudes. However, if you compare the percentage of deep-water radiolarians in the temperate California Current samples to that of the North Pacific gyre, there is a greater percentage of deep-water radiolarians in sediments under the California Current. This difference related to the upwelling that occurs in the California Current, which may carry deep-water radiolarians toward the surface, where they die and rain back down on the sea floor. Alternatively, there may be a larger standing crop of deep-water radiolarians under such high-productivity regions, as noted for other deep-dwelling organisms (see Exercise 20). One other difference between the samples is a greater relative abundance of collosphaerid radiolarians in

the oligotrophic North Pacific gyre (Figure 23-4a) compared to the other sites. These radiolarians form colonies with algae. The radiolarian-alga association appears to be mutually beneficial (symbiotic), allowing high standing crops of collosphaerids in temperate, oligotrophic surface waters, and thereby high skeletal deposition rates on the sea floor below.

Such modern ecological and biological data on living microorganisms serve as a guide for reconstructing the ancient environmental conditions that prevailed when their fossil counterparts were forming sea-floor sediments. Other approaches to reconstructing past conditions include analyzing the chemistry of the skeletons and relating the shape of individual members of a species to their growth environment; you will have an opportunity to do the latter in the exercises that follow.

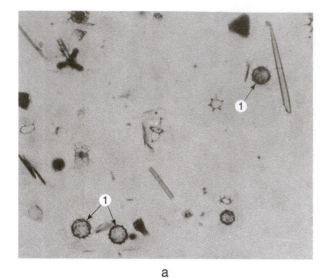

a

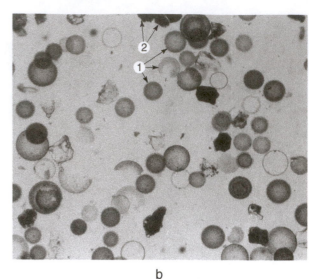

b

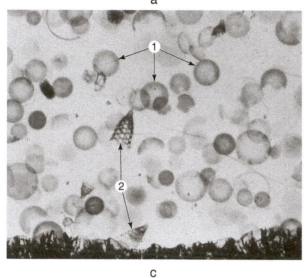

c

Figure 23-4 Radiolarian assemblages from temperate waters. (a) An oligotrophic assemblage from the middle of the North Pacific gyre, showing high diversity and predominance of collosphaerid radiolarians. (b) A eutrophic assemblage from the California Current region, exhibiting low diversity, abundant diatoms (1), and terrigenous sediment (2). (c) Another eutrophic assemblage from the California Current region, with low diversity and abundant diatoms (1), but also containing abundant deepwater radiolarians (2), indicating that the sample was collected from a region of more intense upwelling than the assemblage in (b).

DEFINITIONS

Biostratigraphy. The use of the vertical distribution of fossil species in the rock record to determine the relative ages of different rock units.

Fossil. The preserved remains of a once living organism.

Range. The interval of time during which a particular fossil species lived on earth.

Relative time scale. A biostratigraphy-based scale that divides up the history of the earth into distinct time intervals. Each time interval is defined by the presence of specific fossil species.

Exercise 23

Paleoceanography

1. Consider and discuss why the following attributes would make a fossil species particularly good for biostratigraphy.

(a) Morphologically distinct: _____

_____

_____

_____

_____

(b) Abundant: _____

_____

_____

_____

_____

(c) Short geologic range: _____

_____

_____

_____

_____

(d) Wide geographic distribution: _____

_____

_____

_____

(e) Broad environmental tolerance: _____

_____

_____

_____

_____

2. Many species are not good index fossils because they lived within very specific environmental conditions. Although they are not good for biostratigraphy, how might these species be useful to paleoceanographers? _____

_____

_____

_____

_____

_____

_____

_____

_____

_____

3. You are a marine geologist examining the potential for petroleum reservoirs within the sedimentary layers of the offshore region below. Three exploratory cores were collected, and the presence or absence of seven different microfossils (symbolized by α, β, χ, δ, ϵ, ϕ, and γ) was reported for each 1-meter interval of each core. Undersea erosion has taken place in the region and sedimentation rates have varied between the sites over time, so do not assume that different core tops or similar depth intervals represent the same age. Building on the relationships already summarized in the composite range chart, draw in the relative ranges of α, χ, δ, and ϵ using vertical lines.

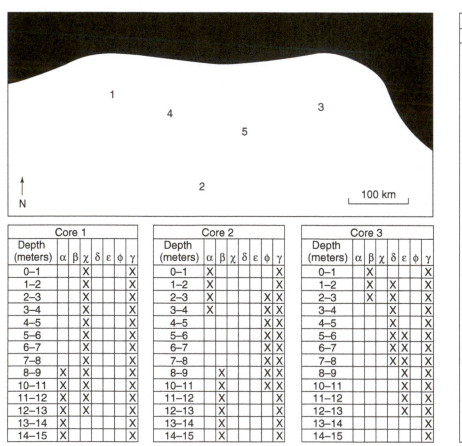

Core 1							
Depth (meters)	α	β	χ	δ	ε	φ	γ
0–1		X					X
1–2		X					X
2–3		X					X
3–4		X					X
4–5		X					X
5–6		X					X
6–7		X					X
7–8		X					X
8–9	X	X					X
10–11	X	X					X
11–12	X	X					X
12–13	X	X					X
13–14	X						X
14–15	X						X

Core 2							
Depth (meters)	α	β	χ	δ	ε	φ	γ
0–1	X						X
1–2	X						X
2–3	X					X	X
3–4	X					X	X
4–5						X	X
5–6						X	X
6–7						X	X
7–8						X	X
8–9			X			X	X
10–11	X					X	X
11–12	X						X
12–13	X						X
13–14	X						X
14–15	X						X

Core 3							
Depth (meters)	α	β	χ	δ	ε	φ	γ
0–1	X						X
1–2	X		X				X
2–3	X		X				X
3–4			X				X
4–5			X				X
5–6			X	X			X
6–7			X	X			X
7–8			X	X			X
8–9				X			X
10–11				X			X
11–12				X			X
12–13				X			X
13–14							X
14–15							X

4. While studying the above exploratory cores, you discover that the sediments deposited during the relative time interval when fossil species β, ϕ, and γ co-occur and a fifth well to a depth where δ, ϵ, and γ co-occur. Using the composite range chart from Question 3, determine if each well is sufficiently deep to "tap" this petroleum, and explain your reasoning. _____

_____

_____

_____

_____

_____

_____

_____

_____

_____

5. Using the data in Figure 23-1, Table 23-1, and the text, what is the geologic epoch and range in millions of years for the radiolarian assemblages illustrated below? Explain the reasoning you used to determine your answer, including why you chose particular species, instead of other species in the assemblage, to date the sample. Also, circle and label examples of each species in the figure. _____

_____

_____

_____

_____

_____

_____

_____

_____

_____

_____

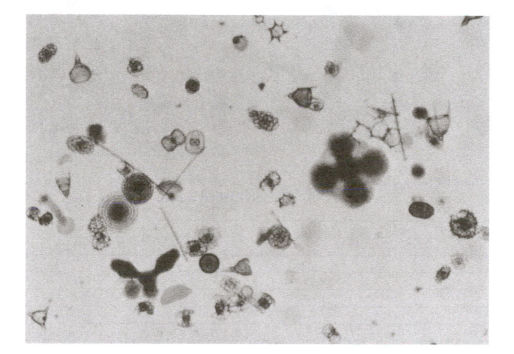

6. Using the ecological data from the text and figures, interpret the paleoceanographic conditions illustrated by the radiolarian assemblage that you dated in Question 5. From what paleoenvironment was it likely deposited, and why? (For example, the sample might record an oligotrophic gyre because it shows high radiolarian diversity and abundant collosphaerids.) _____

_____

_____

_____

_____

_____

7. Date and determine the paleoceanographic conditions of the radiolarian assemblage below, and explain these conclusions as you did for Questions 5 and 6. _____

_____

_____

_____

_____

_____

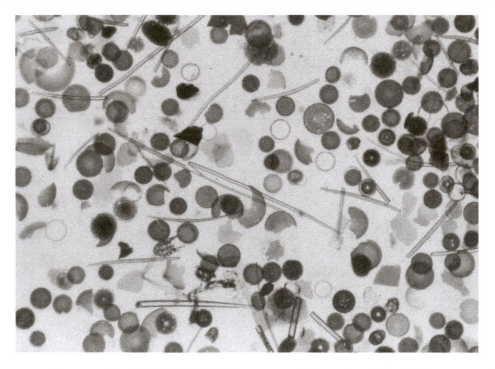

8. Using the composite range chart in Table 23-1, give two possible explanations for how a sample that is late-late Miocene in age and rich in species B, C, D, and E could contain a few species A radiolarians.

_____

_____

_____

_____

_____

_____

_____

9. Using the composite range chart in Table 23-1, explain how during the same time interval (late-late Miocene), one fossil assemblage could contain species B, C, and D, but not E, whereas a fossil assemblage a few hundred kilometers away may contain species B, C, and D as well as E. _____

_____

_____

_____

_____

_____

_____

_____

_____

_____

_____

_____

_____

_____

_____

_____

_____

_____

10. In general, would shallow-dwelling or deep-dwelling radiolarians be better for dating and correlating sedimentary layers across entire ocean basins? Explain your answer. _____

_____

_____

_____

_____

_____

_____

_____

_____

_____

_____

_____

_____

_____

_____

_____

_____

11. In addition to their presence or absence, variations *within* a species may be used to reconstruct ancient environments. The size and shape of individuals of a species are often affected by the environment (e.g., well-nourished children tend to be larger as adults than malnourished children, although genetics also plays a role). Dr. James Kennett, a paleoceanographer, examined shape variation in individuals of the planktonic foraminiferan species *Globorotalia truncatulinoides* and found that the ratio of width to height in the species is highly correlated with the seawater temperature in which it lives (see graph, below). Thus, this ratio may be used to reconstruct local surface temperature variations. Below you are provided a summary of sample ages and average height : width ratios from a core in the North Atlantic. Using the regression equation

$$\text{Temperature} = -55.9 \ (\text{average height:width}) + 97.9$$

convert the given ratios into paleotemperatures and plot them according to their ^{14}C-based age in the blank graph.

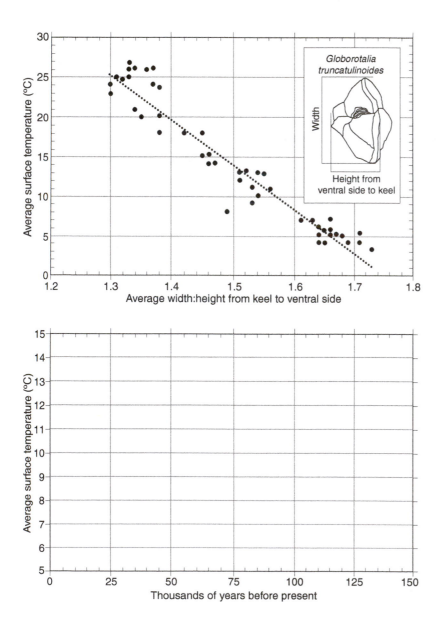

Age (kya)	Average Width:Height	Temperature (°C)
0	1.50	
2	1.51	
9	1.55	
13	1.60	
19	1.64	
27	1.62	
32	1.61	
41	1.61	
53	1.59	
59	1.60	
64	1.61	
70	1.58	
81	1.54	
85	1.56	
92	1.56	
99	1.54	
104	1.55	
109	1.56	
110	1.56	
115	1.53	
122	1.50	
127	1.55	
130	1.59	
146	1.62	

(a) When do temperatures appear to have been the warmest in the region? _____

_____

_____

_____

(b) Based on numerous, independent types of data, the last glacial age occurred about 18 kya (thousand years ago). Is your local data consistent with this general observation? _____

_____

_____

_____

(c) Over the last 1.8 million years, the earth has oscillated between ages termed glacial (relatively cold, more polar ice) and interglacial (relatively warm, less polar ice). According to your data, when (in kya) was there an interglacial most similar to modern conditions? _____

_____

_____

_____

_____

_____

(d) Do there appear to be any other, perhaps less pronounced, shifts between glacial and interglacial ages? _____

_____

_____

_____

_____

When do these transitions occur? _____

_____

_____

_____

_____

(e) Based on your data, which seem to end more rapidly: glacial or interglacial ages? _____
Upon what evidence do you base this answer? Hypothesize what processes might favor this rapidity.

_____

_____

_____

_____

_____

_____

_____

_____

_____

_____

Conversion Factors and Numerical Data

CONVERSION FACTORS

Length

1 micrometer (μm)	0.001 millimeter
1 millimeter (mm)	1000 micrometers
	0.1 centimeter
	0.001 meter
1 centimeter (cm)	10 millimeters
	0.394 inch
	10,000 micrometers
1 meter (m)	100 centimeters
	39.4 inches
	3.28 feet
	1.09 yards
1 kilometer (km)	1000 meters
	1093 yards
	3280 feet
	0.62 statute mile
	0.54 nautical mile
1 inch (in.)	25.4 millimeters
	2.54 centimeters
1 foot (ft)	30.5 centimeters
	0.305 meters
1 yard	3 feet
	0.91 meter
1 fathom	6 feet
	2 yards
	1.83 meters
1 statute mile	5280 feet
	1760 yards
	1609 meters
	1.609 kilometers
	0.87 nautical mile
1 nautical mile	6076 feet
	2025 yards
	1852 meters
	1.15 statute miles

Speed

1 statute mile per hour	1.61 kilometers per hour
	0.87 knot
1 knot	1.15 miles per hour
	1.85 kilometer per hour
1 kilometer per hour	0.62 mile per hour
	0.54 knot
Velocity of sound in sea water at 34.85 parts per thousand (‰)	4945 feet per second
	1507 meters per second
	824 fathoms per second

Mass

1 kilogram (kg)	2.2 pounds
	1000 grams
1 metric ton	2205 pounds
	1000 kilograms
	1.1 tons
1 pound	16 ounces
	454 grams
	0.45 kilogram
1 ton	2000 pounds
	907.2 kilograms
	0.91 metric ton

Pressure

1 atmosphere (sea level)	14.7 pounds per square inch
	33.9 feet of fresh water
	29.9 inches of mercury
	33 feet of seawater

(*continued*)

CONVERSION FACTORS (*continued*)

Time

1 hour	3600 seconds
1 day	24 hours
	1440 minutes
	86,400 seconds
1 calendar year	31,536,000 seconds
	525,600 minutes
	8760 hours
	365 days

Temperature

$$°C = \frac{°F - 32}{1.8}$$ (convert °F to °C)

$$°F = (1.8 \times °C) + 32$$ (convert °C to °F)

100°C = 212°F	(water boils)
40°C = 104°F	(uncomfortably hot)
30°C = 86°F	(hot day)
20°C = 68°F	(perfect day)
10°C = 50°F	(brisk fall day)
0°C = 32°F	(water freezes)

Volume

1 cubic inch (in.^{3})	16.4 cubic centimeters
1 cubic foot (ft^3)	1728 cubic inches, 28.32 liters, or 7.48 gallons
1 cubic centimeter (cc; cm^3)	0.061 cubic inch
1 liter	1000 cubic centimeters, 61 cubic inches, 1.06 quarts, or 0.264 gallon
1 cubic meter (m^3)	10^6 cubic centimeters, 264.2 gallons, or 1000 liters
1 cubic kilometer (km^3)	10^9 cubic meters, 10^{15} cubic centimeters, or 0.24 cubic mile

Multiples

Exponent	Value	Name	Prefix
10^{12}	1,000,000,000,000	trillion	tera
10^9	1,000,000,000	billion	giga
10^6	1,000,000	million	mega
10^3	1,000	thousand	kilo
10^2	100	hundred	hecto
10^1	10	ten	deka
10^{-1}	0.1	tenth	deci
10^{-2}	0.01	hundredth	centi
10^{-3}	0.001	thousandth	milli
10^{-6}	0.000001	millionth	micro
10^{-9}	0.000000001	billionth	nano
10^{-12}	0.000000000001	trillion	pico

Area

1 square inch (in.^{2})	6.45 square centimeters
1 square foot (ft^2)	144 square inches
1 square centimeter (cm^2)	0.155 square inch or 100 square millimeters
1 square meter (m^2)	10^4 square centimeters or 10.8 square feet
1 square kilometer (km^2)	247.1 acres, 0.386 square mile, or 0.292 square nautical mile

NUMERICAL DATA

Scaling Ratios for Charts

Number of nautical miles per inch on the chart	Reciprocal of the natural scale divided by 72,913
Number of statute miles per inch on the chart	Reciprocal of the natural scale divided by 63,360
Number of inches on the chart per nautical mile	Natural scale times 72,913
Number of inches on the chart per statute mile	Natural scale times 63,360

Equivalences in Concentration of Seawater

Seawater with 35 grams of salt per kilogram of seawater	3.5 percent 35.00 parts per thousand (‰) 35,000 parts per million (ppm)

Area, Volume, and Depth of the World's Oceans

Body of water	Area (10^6 km^2)	Volume (10^6 km^3)	Mean depth (m)
Atlantic Ocean	82.4	323.6	3926
Pacific Ocean	165.2	707.6	4284
Indian Ocean	73.4	291.0	3963
All oceans and seas	361.0	1370.0	3795

Appendix **B**

The Geologic Time Scale

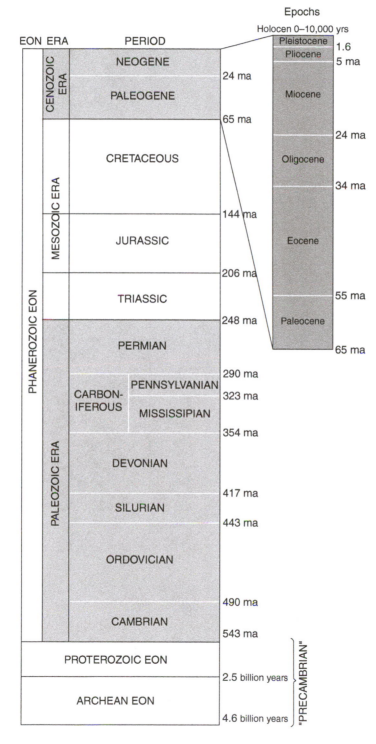

Epochs

Holocen 0–10,000 yrs

EON	ERA	PERIOD		Epochs	
	CENOZOIC ERA	NEOGENE		Pleistocene	1.6
				Pliocene	5 ma
			24 ma	Miocene	
		PALEOGENE			
			65 ma		24 ma
	MESOZOIC ERA	CRETACEOUS		Oligocene	
					34 ma
			144 ma	Eocene	
		JURASSIC			
			206 ma		55 ma
		TRIASSIC		Paleocene	
			248 ma		65 ma
	PALEOZOIC ERA	PERMIAN			
			290 ma		
		CARBON-IFEROUS — PENNSYLVANIAN	323 ma		
		CARBON-IFEROUS — MISSISSIPIAN	354 ma		
		DEVONIAN			
			417 ma		
		SILURIAN	443 ma		
		ORDOVICIAN			
			490 ma		
		CAMBRIAN			
			543 ma		
PROTEROZOIC EON			2.5 billion years		
ARCHEAN EON			4.6 billion years		

PHANEROZOIC EON

"PRECAMBRIAN"

Bibliography

General References

Atwater, B. F., and others, *Surviving a Tsunami—Lessons from Chile, Hawaii, and Japan,* U.S. Geological Survey Circular 1187, 1999, 19 pp.

Aubrey, D. G., "Coastal Erosions Influencing Factors Include Development, Dams, Wells, and Climate Change," *Oceanus,* 1997, v. 36, n. 2, pp. 5–9.

Bascom, W., *Waves and Beaches.* New York: Anchor Books, Doubleday, 1964, 267 pp.

Davidson, K., "What's Wrong with the Weather?" El Niño Strikes Again, *Earth,* 1995, v. 4, n. 3, pp. 24–33.

Davis, R., *The Evolving Coast,* Scientific American Library. New York: W. H. Freeman and Co., 1996, 233 pp.

Garrison, T., *Oceanography.* Belmont, California: Wadsworth Pub. Co., 1999.

Gross, M. G., *Oceanography.* Englewood Cliffs, New Jersey: Prentice-Hall, 1993, 446 pp.

Kennett, J. P., "*Globorotalia truncatulinoides* as a paleo-ceanographic index," *Science,* 159: 1461–1462, 1968.

Mero, J. L., *The Mineral Resources of the Sea.* New York: Elsevier, 1965, 312 pp.

Moores, E. M. (Ed.), *Shaping the Earth—Tectonics of Continents and Oceans.* New York: W. H. Freeman and Company, 1990.

Parfit, Michael, "Living With Natural Hazards." *National Geographic,* 1998, January.

Pinet, P. R., 1992, *Oceanography—An Introduction to the Planet Oceanus.* St. Paul, Minnesota: West Publishing Co.

Pipkin, B. W., and Trent, D. D., *Geology and The Environment.* Belmont, California: Brooks/Cole Publishers, 2000, 463 pp.

Ricketts, E. F., and Calvin, J., *Between Pacific Tides,* 4th ed. Stanford, California: Stanford University Press, 1968, 614 pp.

Sverdrup, H. U., Johnson, M. W., and Fleming, R. H., *The Oceans, Their Physics, Chemistry and General Biology.* Englewood Cliffs, New Jersey: Prentice-Hall, 1942, 1087 pp.

Williams, S. J., Dodd, K., and Gohn, K. K., *Coasts in Crisis,* U.S. Geological Survey Circular 1075, 1991.

Scientific American References

Anderson, D. M., "Red Tides," *Scientific American,* August 1994.

Bailey, H. S., Jr., "The Voyage of the *Challenger,*" *Scientific American,* May 1953.

Bascom, W., "Beaches," *Scientific American,* August 1960. Offprint No. 845, W. H. Freeman and Company, New York.

Bascom, W., "Ocean Waves," *Scientific American,* August 1959. Offprint No. 828, W. H. Freeman and Company, New York.

Beardsley, T., "Death in the Deep," *Scientific American,* November 1998.

Bonatti, E. "The Earth's Mantle Below the Oceans," *Scientific American,* March 1994.

Boyd, C. and Clay, J., "Shrimp Aquaculture and the Environment," *Scientific American,* June 1998.

Broecker, W. S., "Chaotic Climate," *Scientific American,* November 1995.

Brown, B. E. and Ogden, J. C., "Coral Bleaching," *Scientific American,* January 1993.

Courtillot, V., and Vink, B., "How Continents Break Up," *Scientific American,* January 1983.

Dalziel, I. W. D., "Earth before Pangea," *Scientific American,* January 1995.

Dietz, R. W., and Holden, J. C., "The Break-Up of Pangaea," *Scientific American*, October, 1970.

Dolan, R., and Lins, H., "Beaches and Barrier Islands," *Scientific American*, January 1987.

Emery, K. O., "The Continental Shelves," *Scientific American*, September 1969. Offprint No. 882, W. H. Freeman and Company, New York.

Gonzalez, F., "Tsunami," *Scientific American,* May 1999.

Heezen, B. C., "The Origin of Submarine Canyons," *Scientific American*, August, 1956.

Hurley, P. M., "The Confirmation of Continental Drift," *Scientific American*, April 1968. Offprint No. 874, W. H. Freeman and Company, New York.

Kort, V. G., "The Antarctic Ocean," *Scientific American*, September, 1965. Offprint No. 860, W. H. Freeman and Company, New York.

Larson, R. L., "The Mid-Cretaceous Superplume Episode," *Scientific American,* February 1995.

Lynch, D., "Tidal Bores," *Scientific American*, April 1982.

MacIntyre, F., "Why the Sea Is Salt," *Scientific American*, November 1970. Offprint No. 893, W. H. Freeman and Company, New York.

Mack, W. N., and Leistikow, E. A., "Sands of the World," *Scientific American*, August 1996.

Munk, W., "The Circulation of Oceans," *Scientific American*, September 1995. Offprint No. 813, W. H. Freeman and Company, New York.

Murphy, J. B., and Nance, R. D., "Mountain Belts and the Supercontinent Cycle," *Scientific American*, April 1992.

Murphy, R. C., "The Oceanic Life of the Antarctic," *Scientific American*, September 1962.

Pratson, L. F., and Haxby, W. F., "Panoramas of the Seafloor," *Scientific American*, June 1997.

Safina, C., "The World's Imperiled Fish," *Scientific American*, November 1995.

Schmitt, R. W., Jr., "The Ocean's Salt Fingers," *Scientific American*, May 1995.

Schneider, D., 1997, "Hot Spotting," *Scientific American*, April 1997.

Schneider, D., 1997, "Rising Seas," *Scientific American*, April, 1997.

Stommel, H., "The Anatomy of the Atlantic," *Scientific American*, September 1969. Offprint No. 810, W. H. Freeman and Company, New York.

Wiebe, P., "Rings of the Gulf Stream," *Scientific American*, March 1983.

World-Wide-Web Sites

El Niño

http://www.ogp.noaa.gov/enso/
http://www.elnino.noaa.gov
http://www.pmel.noaa.gov/toga-tao/
http://www.coaps.fsu.edu/lib/enso_sites.html

Hurricanes

http://www.nhc.noaa.gov
http://www.nodc.noaa.gov
http://www.hurricanes98.com/

La Niña

http://www.pmel.noaa.gov/toga-tao/

U.S. Geological Survey

http://www.usgs.gov

Weather (severe)

http://www.nws.noaa.gov

Weather (tropical Atlantic)

http://www.goes.noaa.gov

Year of Oceans

http://www.yoto.com

Ocean Productivity

http://seawifs.gsfc.nasa.gov/SEAWIFS/IMAGES/
biosphere_160east_ann.gif
http://seawifs.gsfc.nasa.gov/SEAWIFS.html

Plate Tectonics

http://www.seismo.unr.edu/ftp/pub/louie/class/
100/plate-tectonics.html

Remote Sensing of Oceans

http://www.ssec.wisc.edu/data/
http://www.nodc.noaa.gov/General/satellite.html

Seaweb

http://www.seaweb.org

Tsunami

http://www.pmel.noaa.gov/tsunami-hazard
http://www.usc.edu/dept/tsunamis